Polymers:
Structure and Properties

Polymers:

STRUCTURE AND PROPERTIES

C. A. Daniels, Ph.D., P.E.
Senior Materials Engineer
Affiliated Consulting Engineers

TECHNOMIC PUBLISHING CO., INC.
LANCASTER · BASEL

Published in the Western Hemisphere by
Technomic Publishing Company, Inc.
851 New Holland Avenue
Box 3535
Lancaster, Pennsylvania 17604 U.S.A.

Distributed in the Rest of the World by
Technomic Publishing AG

Printed in the United States of America
10 9 8 7 6 5 4 3 2

Main entry under title:
Polymers: Structure and Properties

A Technomic Publishing Company book
Bibliography: p. 103
Index: p. 105

Library of Congress Card No. 88-50948
ISBN No. 87762-552-2

TABLE OF CONTENTS

INTRODUCTION

This book is intended to be used by practicing engineers and scientists and those who would like to be. It is organized with a minimum of repetition and each segment is designed to be self-contained. Thus, it can easily be used as a reference book, although it was written with the approach common to most technical texts in that the latter chapters build upon material presented in the earlier.

In order to streamline the material presented, the text refers only to high molecular weight polymers, usually considered to be structural polymers. This information is also generally applicable to low molecular weight materials such as paints and solvents with the clear exception of mechanical properties. Foams, due to their commercial importance, are given separate treatment where necessary.

The book has three major divisions: the text, the data section, and the glossary. The text begins with a discussion of structure and bonding within a polymer. It is followed by a chapter on additives. This chapter attempts to deal with the endless possibilities that exist when a pure polymer is modified. The effects of each type of additive are given here. All subsequent discussions of physical and mechanical properties assume a pure polymer, except the elements present as a result of cross-linking in rubbers. In this way, the discussions can be more specific and the reader can adjust the reasoning to include any additives under consideration. Similarly, the data section only contains data for the pure polymer, again with the addition of cross-linking agents in rubbers. All commercially important polymers are listed here with available data. A listing for the general or average properties is also given. The glossary defines terms used in the text. These are usually general terms that have more specific meanings for polymers. Only the usage relative to polymers is given to make the distinction more obvious. Glossary terms appearing in the text are set in small caps.

The illustrations in Chapter 1 may seem unusual at first glance since the atoms are not depicted as the expected round balls. In polymers the bond angle is highly influenced by the *p* orbital which is partly concealed by the round *s* orbital. Thus, the *s* orbital has been reduced to a dotted line so that the *p* orbital may be more apparent. Furthermore, only pure bonds will be depicted although the bonds will tend to smear. Each illustration of this type is drafted to the same scale so that direct comparisons can be made.

Structure and Bonding 1

BONDING WITHIN THE POLYMER

The most common interaction between atoms within a POLYMER MOLECULE is the COVALENT BOND. This type of CHEMICAL BONDING not only produces a stable bond of high strength, but also allows the entire polymer molecule to remain uniformly strong with each additional bond added to it. Thus the very large molecules that comprise a structural POLYMER are as resistant to bond breakage as the smaller molecules from which they were built.

The electrons that form the covalent bond are in the OUTER SHELL of the bonding atom and are either *s* or *p* type. These *s* and *p* electrons have distinctly different ORBITALS as can be appreciated in Figure 1.1. Since bonding stability is dependent on the maximum overlap of the bonding electron orbitals, the bonding of *s* and *p* electrons differs substantially. The *s* electron's spherical orbital allows bonding in any direction and is also more compact than the *p* orbital associated with the same energy shell. Thus, a bond between two *s* electrons, although non-directional, is inherently weaker than any other combination because it has the least overlap. The next stronger bond would be between the *s* electron of one atom and the *p* electron of another atom. Such a bond will have its direction dictated by the position of the *p* ORBITAL. The strongest combination is between two *p* orbitals. The *p-p* bond also requires the most alignment since both atoms are contributing a directional bond. See Figure 1.2 for an illustration of these overlapping bond conditions.

In addition to single bonds, atoms can produce double and triple bonds with neighbors. When a double bond occurs, the *p* orbitals involved must shift position slightly in order to accommodate simultaneous bonding between the two sets of *p* orbitals. The most common angular position resulting from this shift is about 110°. In addition to shifting, the *p* orbitals must also bond to produce effective overlapping. Figure 1.3 shows such a bent bond configuration of a *p* double bond. From this figure it is clear that a triple bond could easily be accommodated by reducing the bond distance to cause overlapping of the *s* orbitals.

A polymer by its nature requires each central atom of the MONOMER to have at least two bonds, thus allowing for the addition of monomers in a minimum of two directions. With the addition of each bonding possibility comes the reduced possibility that each bond can be of the strongest type since the *p* bonds are directional in nature and only one bond can be associated with each orbital. In order to optimize the bonding of the entire molecule, the bonding types blend to produce a compromise among pure *p-p*, *p-s* and *s-s* types of bonds. The resulting bonds are said to be HYBRIDIZED. Resonating MULTIPLE BONDING is also characteristic of many polymer structures. An example of a resonating multiple bond is shown in Figure 1.4.

Almost all covalent bonds have some element of IONIC BONDING. The bent bond configuration does not increase the amount of ionic character normally present, but the resonating bond and multiple bonds do increase the ionic portion of the bonding. In addition to the subtleties associated with bonding configuration, there are often major components of ionic bonding existing in a polymer bond either due to PENDANT GROUPS that are ionic in nature or due to a BACKBONE that contains dissimilar atoms. An example of the former is the OXYGEN-HYDROGEN (OH) RADICAL, which has a net negative charge due to an extra electron. An example of the latter is the silicon-oxygen (Si-O) backbone bonds of silicates.

BONDING BETWEEN POLYMERIC MOLECULES

All matter is attracted by mass attraction which is, of course, operable on the molecular level as well. Also, all bonds, no matter how covalent, have at least a momentary electron fluctuation which induces a DIPOLE. These two factors are combined in the VAN DER WAALS FORCE, which acts as the cohesive mechanism

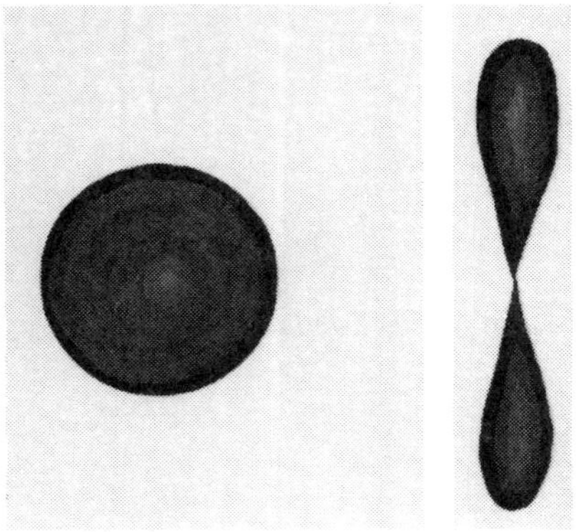

Figure 1.1. SHAPES OF ELECTRON ORBITALS
The spherical *s* orbital on the left can contain up to two electrons. The *p* orbital on the right is just one of the three possible *p* orbitals in the shell. The other two *p* orbitals would be the same shape but located at right angles to this orbital and each other. Each orbital has up to two electrons. Thus, the maximum number of *p* electrons is six per shell.

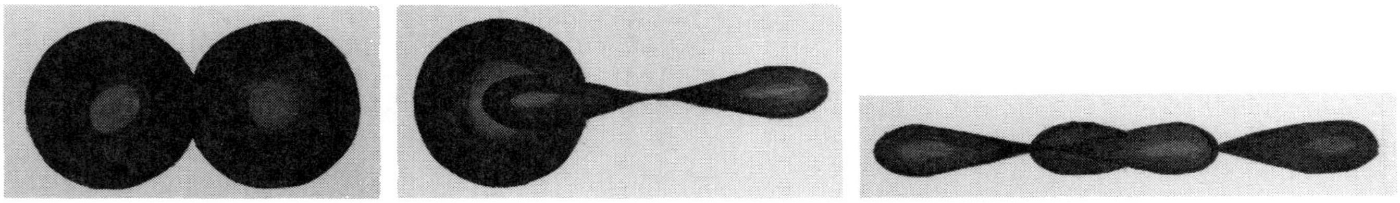

Figure 1.2. CHEMICAL BONDING OF ELECTRONS
The first figure on the left represents the overlap of two *s* orbitals. The middle figure depicts the overlap of an *s* and a *p* orbital. The figure on the right shows two overlapping *p* orbitals. The resulting bonds would blend into one another rather than retain their atomic orbitals.

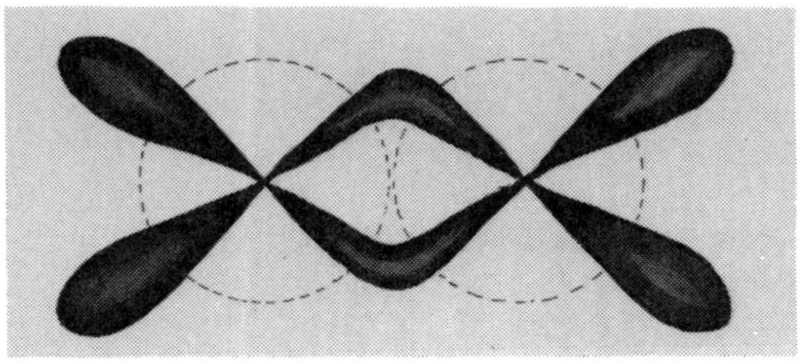

Figure 1.3. DOUBLE BONDING OF ATOMS
The double bonding of two atoms via *p* orbitals is shown above. The *s* orbitals are dotted in for a clearer view of the *p* orbitals. Note that although the two *s* orbitals are very close together, they do not overlap. Also note that the *p* orbitals within an atom are no longer at right angles. The angle they assume is about 110°.

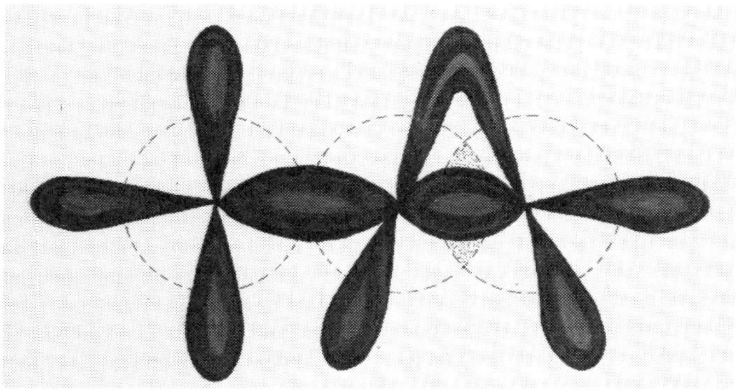

Figure 1.4. TRIPLE BONDING

This schematic shows three atoms with each atom having one *s* and two *p* orbitals in the outer shell. The left atom has a single bond with the center atom which is a *p-p* bond. Note that when the *p* orbitals overlap they form a continuum between the nuclei. The right atom displays a triple bond with the center atom. Since two of them are *p-p* bonds the *p* orbitals take on the 110° angle to each other which results in a bent bond. The *s-s* bond, depicted by the dotted area, is easily accomplished since the higher the number of bonds between two atoms, the shorter the distance between their nuclei. Because the center atom is equally attracted to bonding with both the right atom and the left atom, it can form a triple bond with one and then shift to reform the triple bond with the other, thus resonating the position of the triple bond.

to hold polymeric molecules together. This force is usually about 1% of the covalent bond strength within the polymeric molecule. Obviously, with increased proximity of the molecules, increased size of the molecules, multiple bonds, high ionic character to the backbone bonds and ionic pendants, the value of the van der Waals force will increase and may reach about 10% of the covalent bond strength. Since it is structure-dependent as well as dependent on the chemical composition of the polymer, the strength of a polymer cannot be directly inferred from its chemistry alone.

CONFIGURATION OF THE POLYMER MOLECULE

The majority of polymeric molecules are based on ORGANIC carbon derived from naturally occurring hydrocarbons, which already have some form and structure. Most of them are short chains with repeating units or MONOMERS, with carbon comprising the backbone of the chain and other atoms or groups of atoms satisfying the remaining bonding possibilities. PROCESSING is carried out, to extend the length of the backbone chain. The other bonds may also be modified through processing. The extended chain configuration has proven to have useful properties and as a consequence a great deal of effort has gone into development of procedures through which the resulting chains conform to a specific range of lengths.

When short chains are REACTED to produce longer ones, the reaction is usually controlled so that a LINEAR POLYMER results. This means that the backbone atoms will have another backbone atom joined to them on each side but not in other directions. Having a bond on each side of the backbone does not necessarily result in the backbone being a straight line, since the bonding angles vary with the nature of the bond. The most common form of bonding is the single bond with the *p* orbital being involved. Since this results in a 110° bond for most backbone forming atoms, the resulting backbone has a zig-zag arrangement. This is illustrated in Figure 1.5.

Although the bond angle between backbone atoms is determined by the type and number of bonds, the bonds are often free to rotate in space. When there are bonding atoms or pendants attached to the backbone, the rotation around the backbone usually assumes specific positions due to spatial and repulsive considerations. The most prevalent rotational angle values are 0° or TRANS position, ±180° called CIS position, and ±120° known as GAUCHE position. Figure 1.5 shows backbone atoms in the trans position, Figure 1.6 shows them in the cis position and Figure 1.7 shows them in the gauche position. The occurrence and value of the rotational angles can further be fixed by the presence of side chains or BRANCHED CHAINS attached to the backbone. Branch chains are usually larger than bonded atoms or pendants and thus have a greater effect on the rotational positions of the backbone. These become rather rigid and cannot be easily altered. Thus, the larger the side atom or group of side atoms attached to a backbone, the less freedom the backbone bond has to rotate in space. Figure 1.8 illustrates branching and Figure 1.9 shows a common pendant type. Figure 1.10 illustrates the rotation caused by a pendant to a backbone.

The size alone of an atom bound to a backbone may not determine the amount of rotation that particular atom induces or constrains in the backbone to which it is attached. Repulsive forces due to electronegativity of the atom are superimposed such that the rotational ability of a bound atom may be much greater than just its physical size would suggest. It is also possible that the backbone itself contains structures that dictate the rotational properties of the backbone without consideration for the side structures. Figure 1.11 illustrates a backbone that contains a CARBON RING, which establishes its rotational relationships.

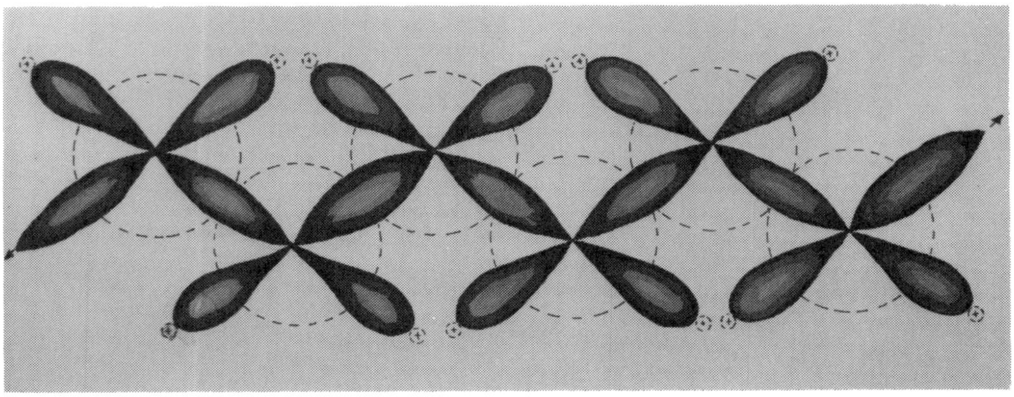

Figure 1.5. A SCHEMATIC OF PART OF A CARBON CHAIN

The arrows show where the chain will continue. Each carbon atom is able to react with another atom at each site which is marked with a + as well as bond to its neighboring carbon atom. An example of this reaction is polyethylene where all of the + would be occupied by hydrogen atoms. The segment of the chain depicted could then be written as:

```
   H  H  H  H  H  H
   |  |  |  |  |  |
 ←-C--C--C--C--C--C-→
   |  |  |  |  |  |
   H  H  H  H  H  H
```

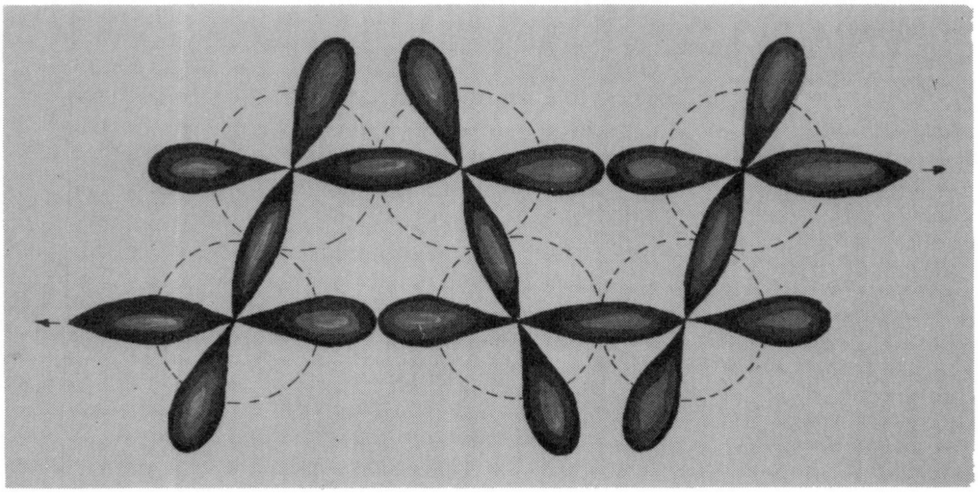

Figure 1.6. THE CIS POSITION IN A CARBON BACKBONE

This schematic shows part of a carbon polymeric chain and the arrows show where the chain will continue. All of the atoms shown are in the cis position, or rotated 180°. Most carbon chains with only single bonds assume very few rotated bonds since the repulsion is greater than in its normal trans position. The potential repulsion can be visualized when comparing the proximity of the unbonded *p* orbital in this figure to the corresponding ones in Figure 1.5 which are in the trans position.

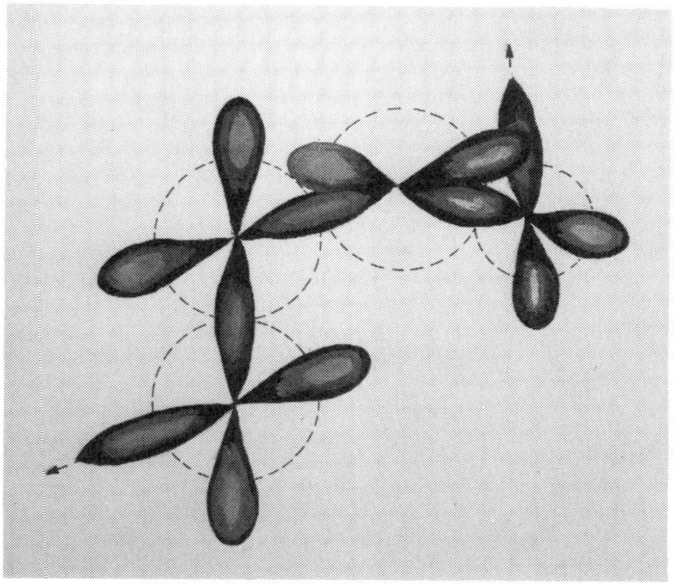

Figure 1.7. THE GAUCHE POSITION FOR A BACKBONE This schematic shows part of a carbon backbone and the arrows indicate where the chain will continue. Going from left to right, the first atom shown is in a trans relationship with the second atom; the second and third atoms are in a gauche, 120° rotation; and the third and fourth atoms are back into a trans relationship. Note that the fourth atom is completely rotated out of the plane of the first two atoms and is "behind" them in space. The third atom is partly in the plane of the first two atoms but its left unbounded *p* orbital is "above" that plane.

Some inorganic materials will also form long chains with proper processing. The backbone element in these is sulfur, tellurium, silicon, selenium, or graphite. In addition, silicon, aluminum or phosphorus alternating with oxygen will form a chain. Figures 1.12 and 1.13 illustrate the silicon-oxygen backbone, which is also typical of the aluminum-oxygen and phosphorus-oxygen backbones. Of the inorganic-based polymers, only the graphite, silicon, and silicon-oxygen backbones have produced materials with ENGINEERING APPLICATION, although a good number of these other types of polymers have been developed and tested with favorable results.

MOLECULAR WEIGHT OF POLYMERS

Depending on the processing procedure used, the molecular weight of the resulting polymer can range from the very lightest of molecules up to huge gels. As the average MOLECULAR WEIGHT increases, the properties continue to change until a plateau is reached. This plateau varies with the property considered and with the chemical composition of the material. However, a useful generality is that the plateau is reached when the number of atoms in the backbone is greater than 1000. If, for example, the monomer CH_2 is considered, it can be seen that the single atom backbone methane is a gas at ambient temperature, whereas octane with eight carbon atoms is a liquid. When the carbon backbone numbers 35, the material has become a grease; by 150 it is a wax and by 500 it has become a hard resin and will retain this character with additional increases in the backbone number. Most polymers that are to be used for load bearing purposes not only have an average backbone number above 1000 but also virtually eliminate the smaller backbones from the distribution of sizes that result from processing. If these small molecules are not eliminated, then at least some of the overall properties of the resulting polymeric material will be altered substantially.

CROSS-LINKING OF POLYMERS

Besides making a polymer of the long chain type it is also possible to make a three-dimensional network. Such networking transforms the polymer into what is basically one huge molecule. The resulting array appears as something between the two-dimensional string of a linear polymer and the regular bonding arrangement of a CRYSTALLINE ceramic. CROSS-LINKED polymers never attain the regular structure of a crystal. Thus, they are always amorphous unless other forces such as an applied load are involved.

The network arrangement in a polymer can be produced by chemically reacting atoms or groups of atoms from one chain with another atom or group of atoms in a neighboring chain. As this reaction proceeds from one site to the next, the polymer becomes more and more cross-linked until all available sites are reacted. This was first done for a commercial product by Goodyear, who introduced sulfur into natural rubber with added heat. The double bonds of the carbon backbone in this VULCANIZING process break preferentially and rebond with the sulfur. The sulfur then serves as a linkage between chains. Of course, the amount of sulfur added limited the amount of reaction which could take place up to the point at which all of the REACTION SITES were used. As the

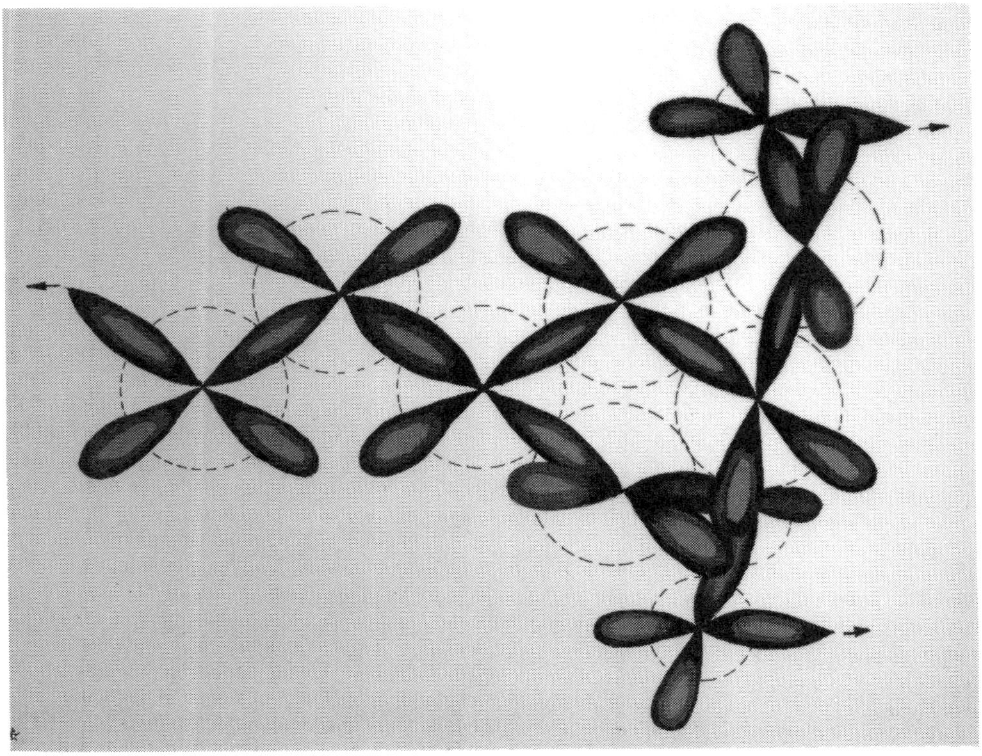

Figure 1.8. BRANCHING OF THE BACKBONE
This is a schematic of part of a carbon backbone; the arrows indicate the continuation of the chains. The third atom from the left is branched such that a new chain comes off of it as well as the normal extension of the chain. The upper part is the "main chain" which shows the next atom in its usual position. The first atom of the branched, or lower section, is shown rotated in the gauche position. A shorthand version is:

```
—trans—trans—trans—trans—cis—gauche—trans—
              |
            gauche
              |
            trans
              |
            trans
```

Also note that the branched atom can only accommodate one more bond. If this were polyethylene the notation would be:

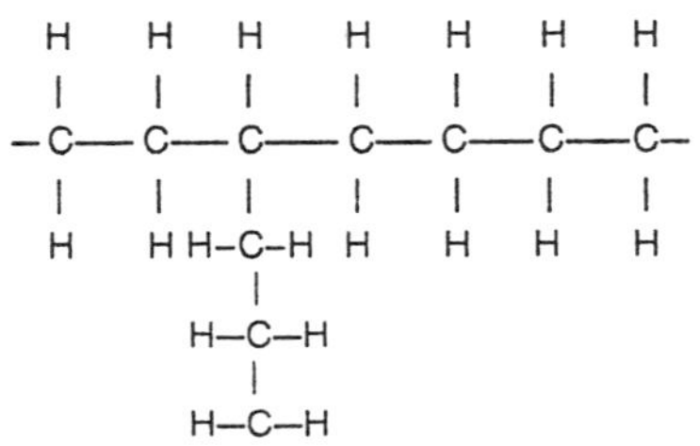

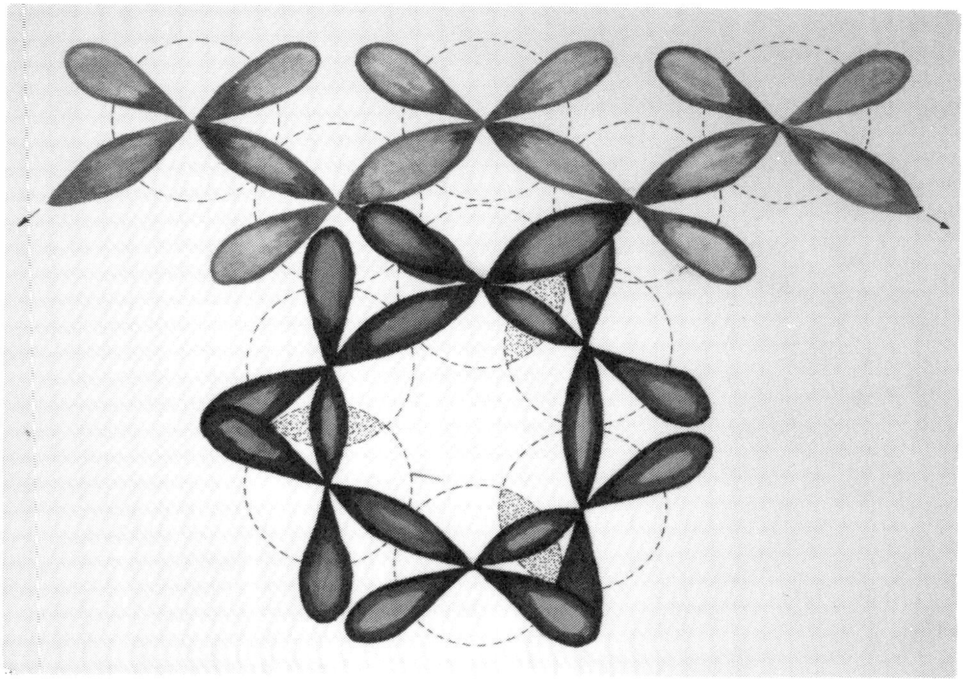

Figure 1.9. A CARBON RING AS A PENDANT

This is a schematic of part of a carbon chain where the arrows indicate the continuation of the chain. The lighter atoms at the top are the normal polymeric chain in the cis formation. The darker atoms represent a carbon ring. Note that the carbon ring has alternating double and single bonds. These double bonds limit the addition of additional atoms to the ring. If this were polyethylene with a plain carbon ring the structure would be:

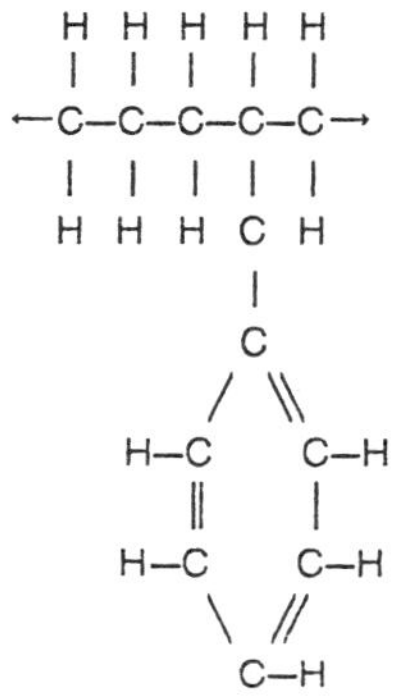

It can also be seen that unbonded p orbitals of the carbon ring overlap at the double bonds. In reality this double bond is a hybrid which is partially s and partially p. Further overlaps can be seen between the pendant and the main chain. These are basically eliminated by rotation of the chain. See Figure 1.10.

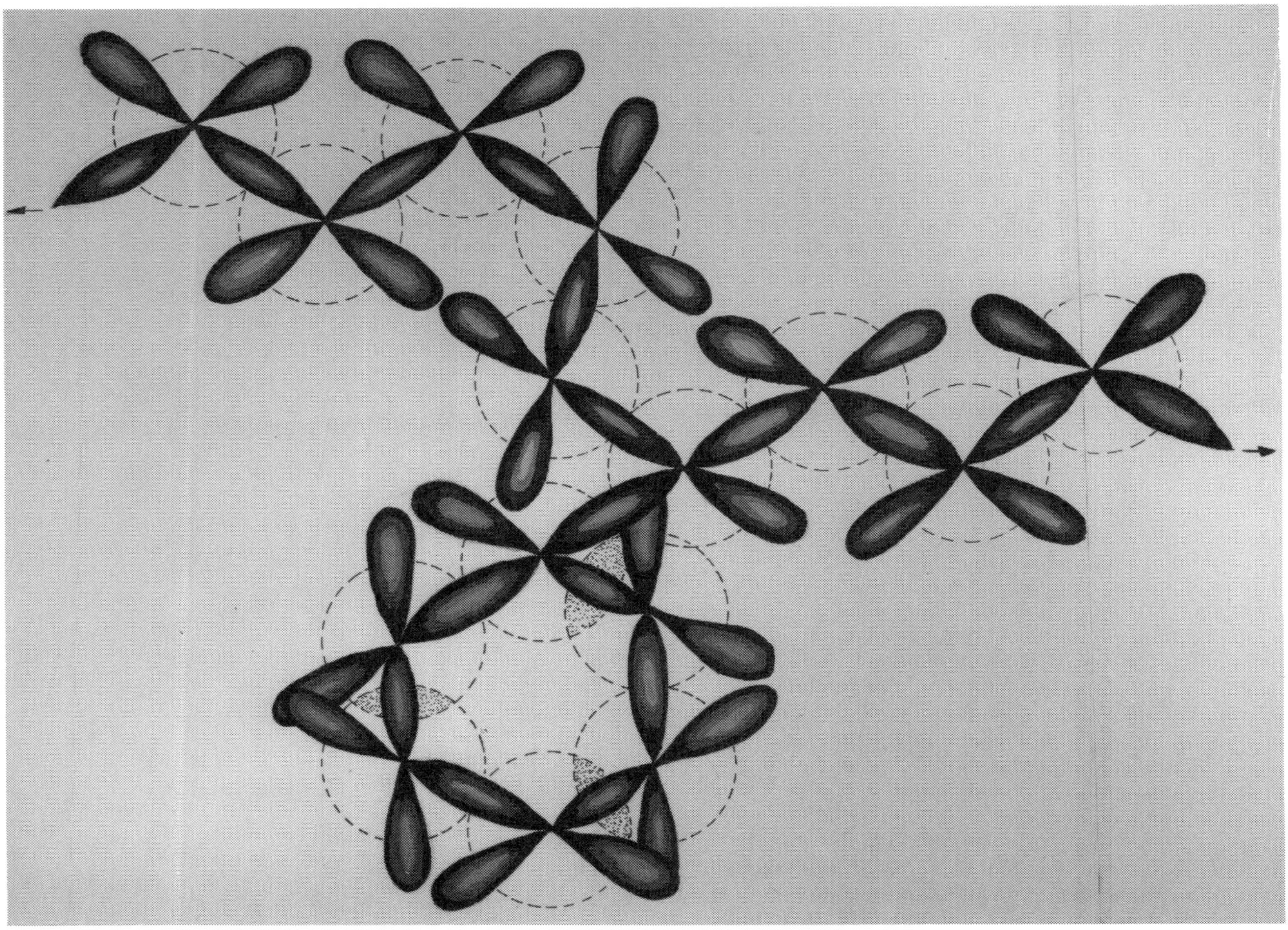

Figure 1.10. ROTATION OF A BACKBONE DUE TO A PENDANT
This schematic shows part of a carbon backbone and the arrows indicate where the chain will continue. The carbon ring at the bottom of the schematic causes the chain to twist in space in order to avoid overlap. If the chain is to remain in the plane of the carbon ring then the twisting is done as follows:

trans–trans–trans–cis–trans–cis–trans–trans–trans–
|
carbon ring

natural rubber went from its native state to increasing cross-linkage, the mechanical properties were correspondingly altered. Thus, the rubber went from a soft gooey mass to a hard tough material. The tailoring of rubber properties by the use of cross-linking remains the primary use of this processing procedure. Diagrams of cross-linked natural rubber can be seen in Figures 1.14 and 1.15.

Another form of cross-linking can be done without adding a linkage atom. This procedure is essentially the making of numerous branches from a central backbone which results in interconnected rings. The number of backbone atoms between branches is fairly large due to the twisting that needs to occur before the backbone is again in a position to reconnect via branching. See Figure 1.8 for a diagram of branching. This form of cross-linking is by its nature limited to about one in every fifty atoms and thus is considered light cross-linking.

FOAMS

Virtually any polymer can be processed with a FOAMING AGENT to form a FOAM. In most cases, a foaming agent in the form of a gas is mixed uniformly throughout the liquid polymer, and the mixture is cooled. What results is usually an OPEN-CELLED material with pentagonal faces and a dodecahedral (honeycomb) structure, although cells of most any size and shape are possible. The polymer phase is fairly uniform in thickness throughout a dodecahedral structure but may collect in the corners. A diagram of this is shown in Figure 1.16.

The properties of the foam may be considered that of the polymer with the addition of a GASEOUS FILLER. As such, most characteristics will be those of the polymer. However, when volume is important, such as in THERMAL CONDUCTIVITY, STIFFNESS FACTOR and floatation, then the foam will act as a composite.

There is no upper limit to the percentage of polymer that can be used relative to the percentage of foaming agent. However, the minimum amount of polymer is usually about 1%. Below 1% the polymer becomes unstable in processing and collapses. It is possible in some polymeric materials to further reduce the polymer percentage by limited cross-linkage. In prac-

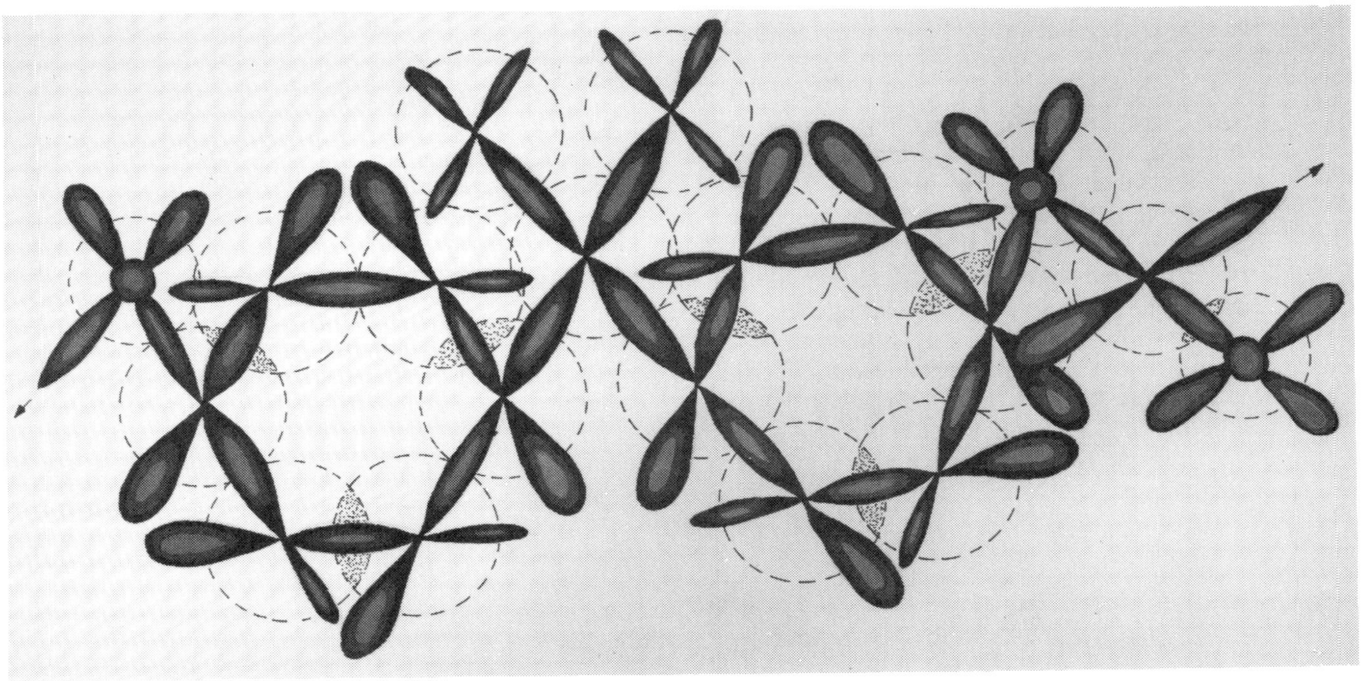

Figure 1.11 AN ILLUSTRATION OF THE MONOMER POLYCARBONATE

In the polycarbonate monomer a carbon ring is part of the backbone. The monomer is:

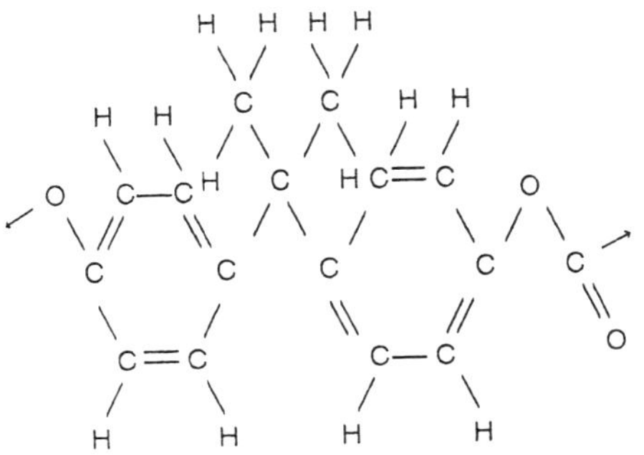

The first atom shown on the left is an oxygen atom which has a *p* orbital at right angles to the plane of the carbon ring, as do the other two oxygen atoms depicted. Wherever the carbon and hydrogen are bound the bond length is actually shorter than the unbound *p* orbital of the carbon atom. This monomer is resistant to rotation, since the orientation of the oxygen-carbon bond fixes the position of the entire monomer. The oxygen and carbon atom on the right side of the monomer appear to overlap. In fact, the oxygen atoms are not in the plane of the ring.

tice this is done both by irradiation and chemical agents. The use of the latter is not favored, although it is usually cheaper, because such lightweight foams are usually designed for medical applications where the chemical agents are an unwanted impurity.

CRYSTALLIZATION

Polymers can conform to a crystalline array, but the difficulties of arranging long chains in three dimensions are such that crystallinity cannot be completely achieved. The degree of crystallinity and the size of the crystals are dependent on how energetically favorable the arrangement is. For polymer chains that have dipoles or are polarizable the intermolecular forces (van der Waals bonds) will encourage close and regular proximity, encouraging crystallinity. On the other hand, if the chain is ATACTIC or has bulky pendant groups, then the spacing between the chains will be large due to the spatial relations of fitting these irregularities into space. The increase in distance between the chains reduces the effectiveness of the intermolecular attracting forces. When this occurs, the polymer is capable of very limited crystallinity. The ability to crystallize does have an effect on the size of the crystals as well. If the polymer has irregularities that make alignment for crystallizing difficult, these irregularities will also impede the number of chains that can be brought into alignment within a space. Thus, polymers which have limited crystallizing ability also tend to produce crystals of small size.

The long chains of a polymer demand more time and mobility to arrange themselves into a crystalline array than other types of materials. As such it is possible to QUENCH a polymer from the melt past the glass TRANSITION TEMPERATURE without producing a crystalline structure. Once a polymer is below its glass transition temperature, the chains do not have sufficient mobility to rearrange for crystallization or any other DIFFUSION-DEPENDENT change. A depression of the molten polymer to approximately nine tenths of the melting point (on the absolute scale) will usually result in the maximum crystallization rate. Once the conditions are favorable for crystallization, the process of nucleation and growth is similar to other crystalline materials. The nucleus usually starts as a small alignment of CHAIN SEGMENTS, which have come together due to random thermal motion. Growth of the nucleus can be encouraged by a directional external force. The force tends to align the chains in its direction, thus helping to order the chains for crystallization. Once crystallization is attained, the polymeric chains are closer together than they would be in their amorphous state. Consequently, the intermolecular attraction between the chains is stronger and the material is more stable. A manifestation of this is the increased melting point of a crystalline polymer com-

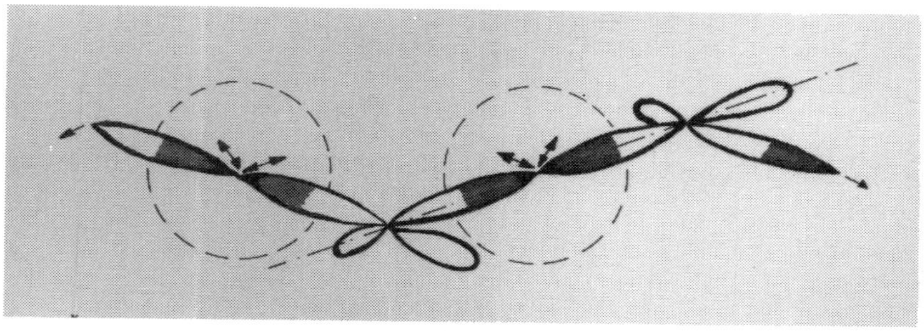

Figure 1.12. THE SILICON-OXYGEN BACKBONE
This is a schematic of the silicon-oxygen backbone of a polymer; the open-end arrows indicate the continuation of the chain; the closed-end arrows show where the silicon bonds to other atoms. The chain shown is:

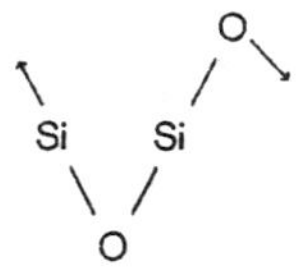

The chain forms a shallow zig-zag with an angle of about 142°. Only the silicon is in the plane of the schematic and thus is painted-in. The oxygen atoms are all shown below the silicon in this schematic. At first glance it may appear that the silicon forms bonds with two oxygens with the same *p* orbital resulting in a 180° bond. If this did happen the oxygen would be in the same plane as the silicon. The arrangement of the bonding *p* orbitals of the silicon with the oxygen can be seen more clearly in Figure 1.13. The section depicted in Figure 1.13 corresponds to a side view of the line designated by —·—·—·—·—

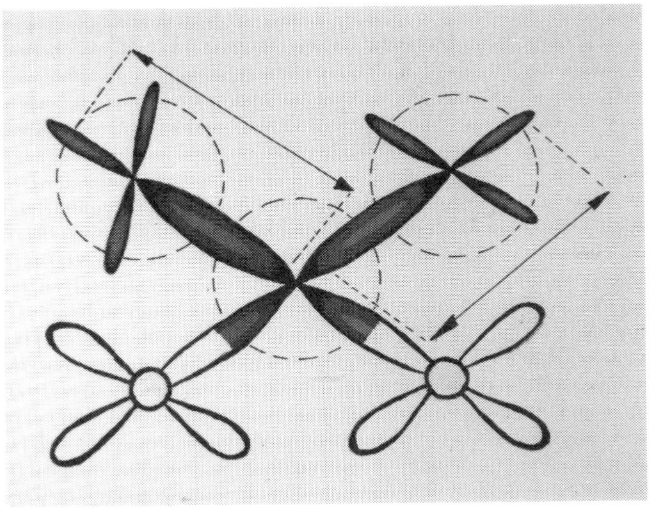

Figure 1.13. SIDE VIEW OF THE SILICON-OXYGEN BACKBONE This schemetic is the side view of Figure 1.12 and is indicated on that figure by the –·–·–·–·– line. Here the double-end arrows have been replaced by CH_3 making this into poly(dimethyl-siloxane). These arrows are still made on this schematic, however, for cross-reference. The monomer shown is:

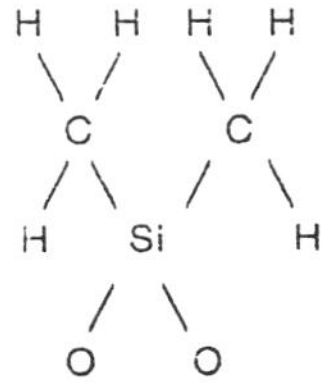

The oxygen atoms have not been painted, in order to indicate that they are in the plane below the silicon. Note that in this arrangement, the silicon *p* orbitals show clearly their characteristic angular relationship of about 110°. The oxygen atom also has about a 110° angle between the two *p* orbitals shown here from the side. The third *p* orbital is shown on its end as a circle. It is this orbital which bonds with the next silicon in the chain.

pared to the same polymer in the amorphous condition. The arrangement of crystalline polyethylene in a UNIT CELL can be seen in Figure 1.17.

LAMELLAR CRYSTALS

Polymer crystals usually form into LAMELLAR structures from which larger SPHERULITES can develop. Lamellae are easily formed from a crystallizable polymer when it is cooled slowly from its melting point. Other very different structures are obtained when polymer crystals are formed by precipitation from a dilute solution or are rapidly cooled from the melt.

The lamellae consist of folded chains, the thickness of which is of the order of 10 nanometers, while the length and width are usually about equal to each other and much greater than the thickness. A single polymeric chain is often about 1000 nm long. Hence, it must fold many times to fit into a lamella. Most chains fold into a single lamella when the cooling is done slowly. Some chains will crystallize into two or more lamellae and form linkages between lamellae; these are TIE MOLECULES. Figure 1.18 illustrates the fold configuration of two lamellae of polyethylene that have tie molecules between them. Figure 1.19 is a detail of Figure 1.18.

FIBRILS

If a polymer is crystallizing from a pure melt, as opposed to forming from a solution, then the lamellae will start to grow from nuclei scattered throughout the melt. Each nucleus will spawn lamellae which will grow radially from the center. The first structure to form from this will be the RADIAL FIBRILS. As the fibrils grow radially outward, they branch to better fill space. Within a fibril the lamellae are aligned and interconnected by tie molecules. Fibrils will usually contain amorphous polymer around the lamellae. Since the fibril is so interconnected and aligned, it acts similarly to a single crystal, although it is composed of many individual crystals or lamellae. Figure 1.20 illustrates part of a fibril.

SPHERULITES

Individual crystals in the form of lamellae are organized into fibrils that are part of the overall structure of a spherulite. The spherulite grows out from the

Figure 1.14. NATURAL RUBBER CROSS-LINKED WITH SULFUR

This schematic shows a section of the natural rubber molecule which has been cross-linked via sulfur to another natural rubber molecule. The open-end arrows indicate where the chains will continue. The arrangement shown is:

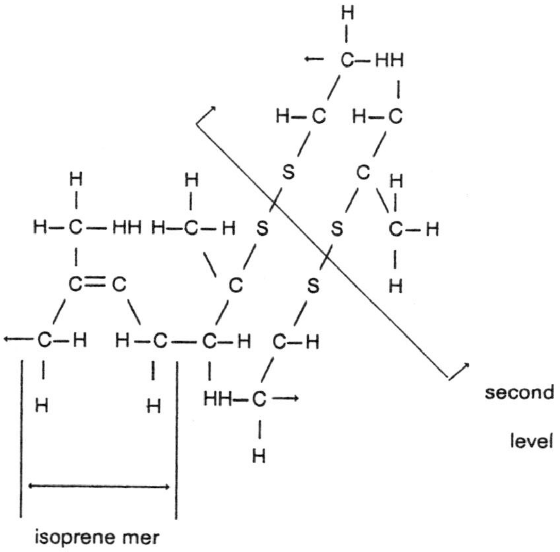

Note that the top part of the schematic (second level) is not painted. This distinguishes the main plane of the schematic from the one above it. The atoms in the main plane of the schematic also have the *s* orbitals dotted in, whereas the second level does not. This schematic represents one of many possible arrangements of cross-linked polyisoprene with each arrangement being quite complex. See Figure 1.15 for the side view of this schematic. The slice taken for the side view is designated by the –·–·–·–·– line.

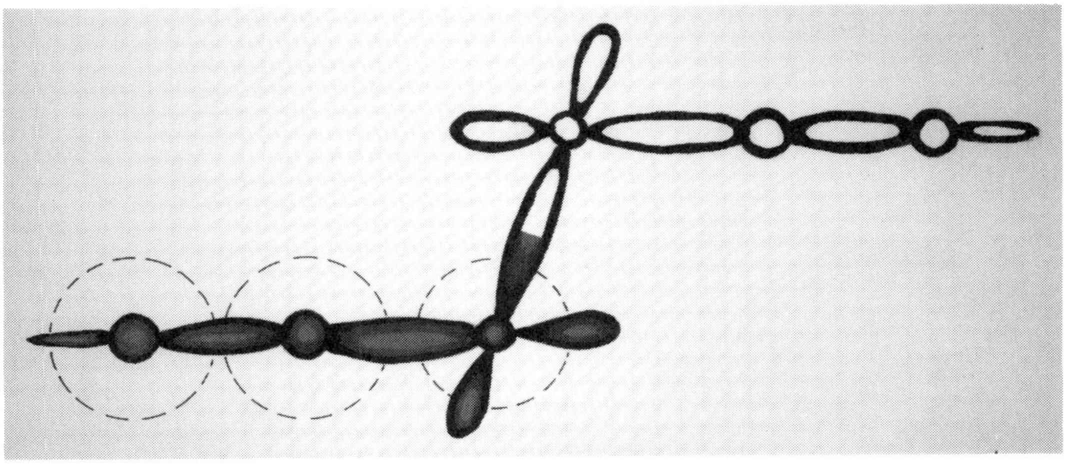

Figure 1.15. CROSS SECTION OF CROSS-LINKED POLYISOPROPRENE
This is a schematic of natural rubber, polyisoproprene, which has been cross-linked with sulfur. See Figure 1.14 for the front view. The arrangement shown is:

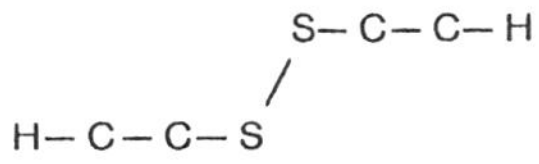

As in Figure 1.14, only the main level is completely painted and the *s* orbitals indicated.

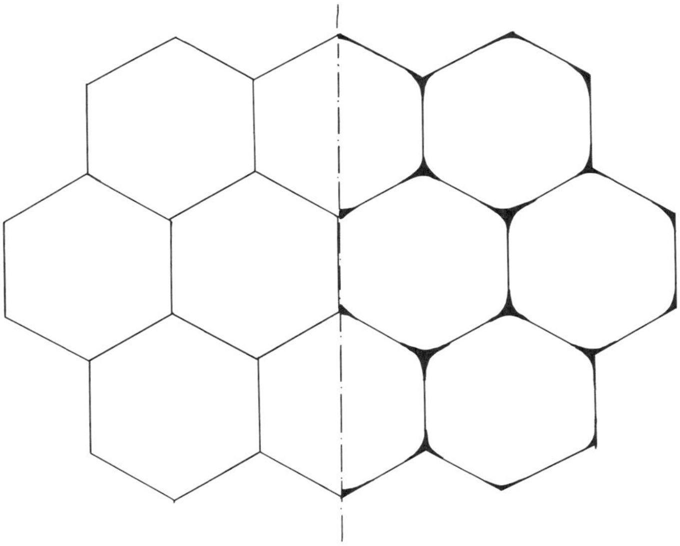

Figure 1.16. FOAM GEOMETRY
This is a diagram of a regular dodecahedron in cross section. Left of the –·–·–·–·– line is the geometric shape of the cells formed by most foams. Right of the –·–·–·–·– line is the same geometric shape with capillary flow to the corners also represented. An open cell simply means that one or more of the sides of the dodecahedra are not completely closed by polymer.

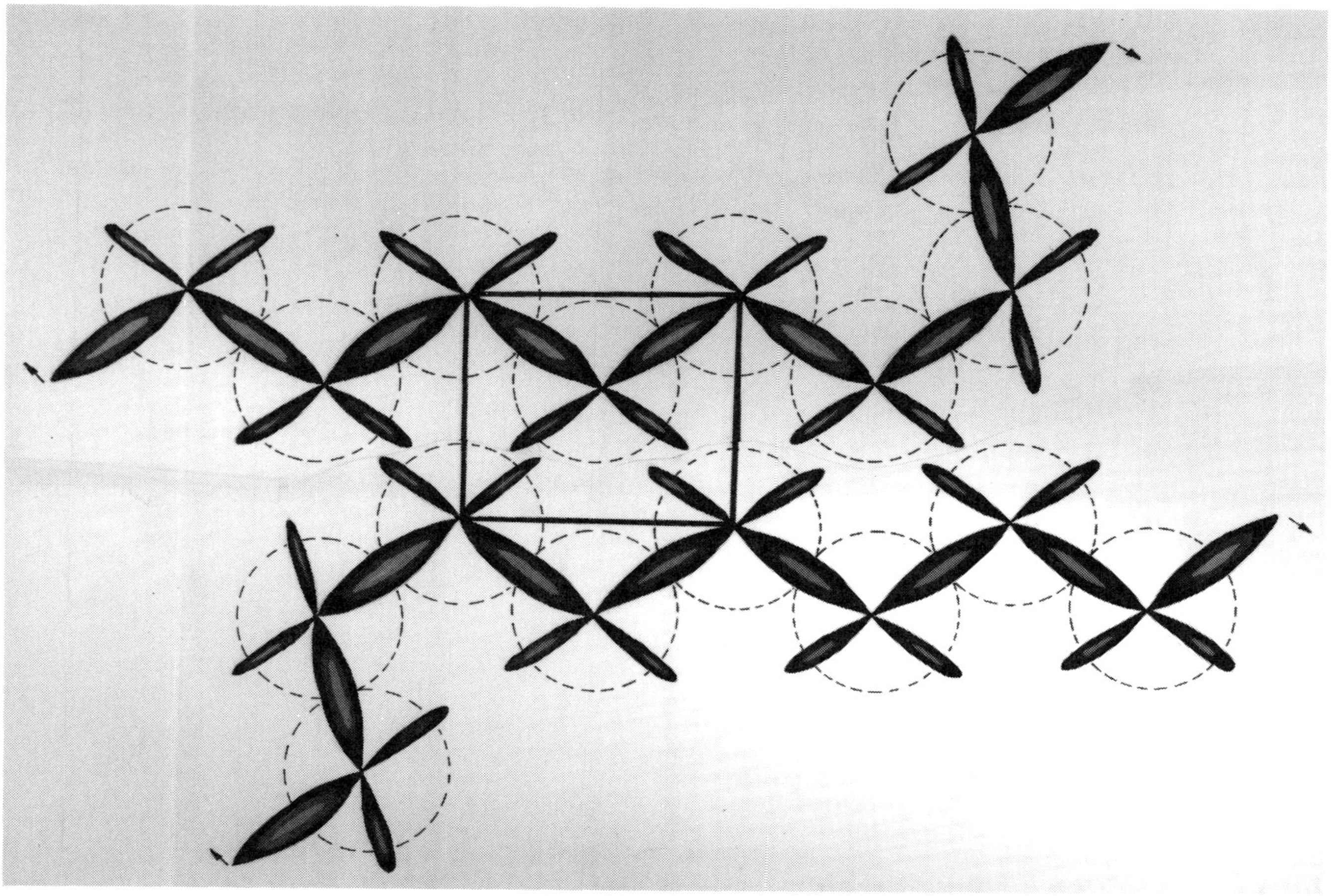

Figure 1.17. CRYSTALLINE UNIT CELL OF POLYETHYLENE
This schematic shows the alignment of polyethylene chains to form part of the unit cell of a crystal. The arrows indicate the continuation of the three chains shown. The center chain is only outlined so as to indicate that it is not in the plane of the other two chains shown. To complete the unit cell (partially marked by the solid straight lines) would require two more chains of polyethylene aligned as the two painted chains are but in a plane farther beyond the outlined chain. The creation of just one unit cell can serve as the nucleation point for the subsequent crystal growth.

center nucleus in a spherically symmetric array of fibrils until it impinges on another spherulite. Like all growing crystal structures, its final size is dependent on the cooling rate of the melt and the nucleation sites of other crystals. As a result, spherulites can range in size from less than 0.001 millimeters to greater than 10 millimeters in diameter. As the fibrils within the spherulite grow, they leave areas within the sphere which are not included as part of any fibril. These areas will provide sites for secondary crystallization and the lamellae that form within them will not, in general, have the same orientation that exists within the fibril. This space will also contain amorphous material. Figure 1.21 shows a diagram of a spherulite.

DEFECT STRUCTURE IN THE CRYSTAL

In addition to the USUAL DEFECTS associated with a crystal made up of atoms or small molecules, a polymeric crystal has defects related to its long chain structure. As the chain folds back and forth upon itself, it can produce irregular fold heights. The chain may also fold back into the lamella at a position other than next to the previous fold of the chain. Most long chains have some amount of branching, side loops, and kinds that may disrupt the regular array of the lamella. All of these irregularities contribute to the instability of the crystal.

Like other types of crystals, polymeric crystals deform in the direction of applied stress. Evidence for this is seen in the elongation of individual spherulites in the direction of stress. Such elongation is accomplished by slip within, as well as between, lamellae with the surrounding amorphous material adjusting accordingly. Applied stress may also bring into alignment chains that were not previously situated favorably for crystallization and produce crystal formation in such a region. As stress is continued, the polymer adjusts until it is unable to continue uniform deformation and NECKING begins. During the deformation of a polymeric crystal, all of the usual defect-producing mechanisms operative in other types of crystals also seem to be in evidence.

In order to alleviate some of the defects introduced by deformation of the crystal, ANNEALING may be done. The principal effect of annealing is to cause an adjustment of the folds of the polymeric chain such that the defects are reduced. In addition, the fold height of the lamellae is often increased by as much as twice its original height. The position of the lamellae also shifts to a lower stress, causing a deterioration in the spherulite organization.

AMORPHOUS STATE

Since there currently exists no experimental procedure that can reveal the short range order that may exist in the amorphous state, any mental images that can be developed about it must be based on other data. Probably the most important information concerning the amorphous state is that it is about 85% as dense as a perfect crystal for most specific polymers. This means that there is a fairly high packing density of polymeric chains in the amorphous state, which would not be achieved merely by random coiling and interacting of the chains—it does not look like a plate of spaghetti. The models that are consistent with the density data suggest that an amorphous polymer has a structure similar to the crystalline polymer, but the regions of ordered crystal-like structure are much smaller than identifiable crystals. There is also a considerably larger volume of material without any indication of ordering.

GLASSY STATE

All amorphous polymers will exhibit a GLASSY region and some will have an exclusively glassy state. In the glassy state thermal energy is not sufficient to permit even segments of the polymer chains to move relative to one another. Thus it is an essentially frozen structure in which thermal motion is limited to vibra-

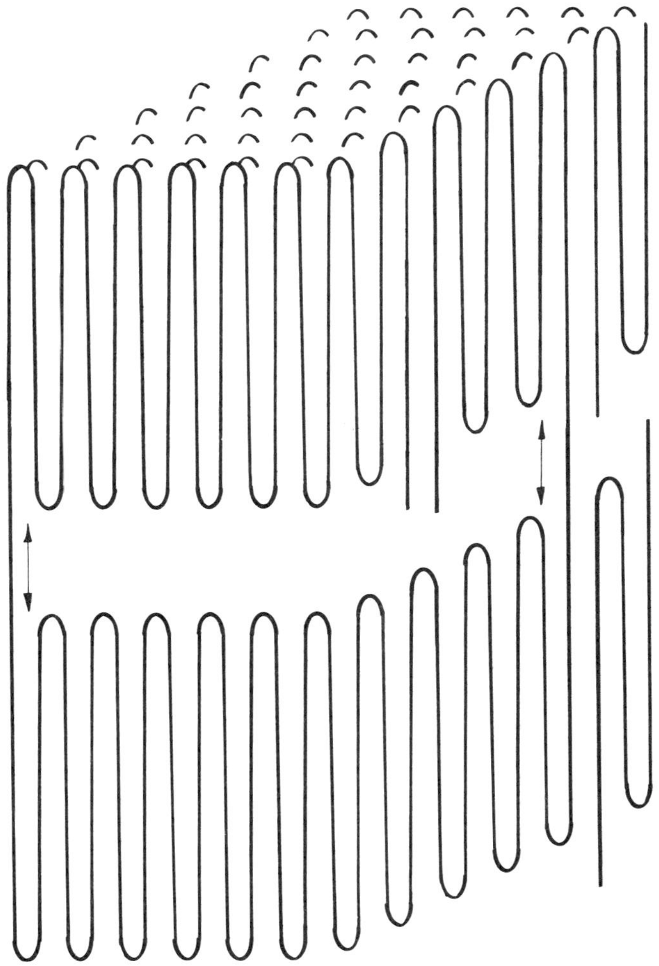

Figure 1.18. LAMELLAE FORMED IN POLYETHYLENE
This is an illustration of the lamellae formed by polyethylene. The upper lamella also shows the tops of chains on the top surface of the lamella. Only two sides of the lower lamella are shown to illustrate tie molecules, which are marked by parallel double-headed arrows in the region where they "tie" the two lamellae together. Note that both sides of the lamellae shown have folded surfaces rather than just one side as might be expected. The folds face the side to which they are folded, rotating out from the center of the lamella.

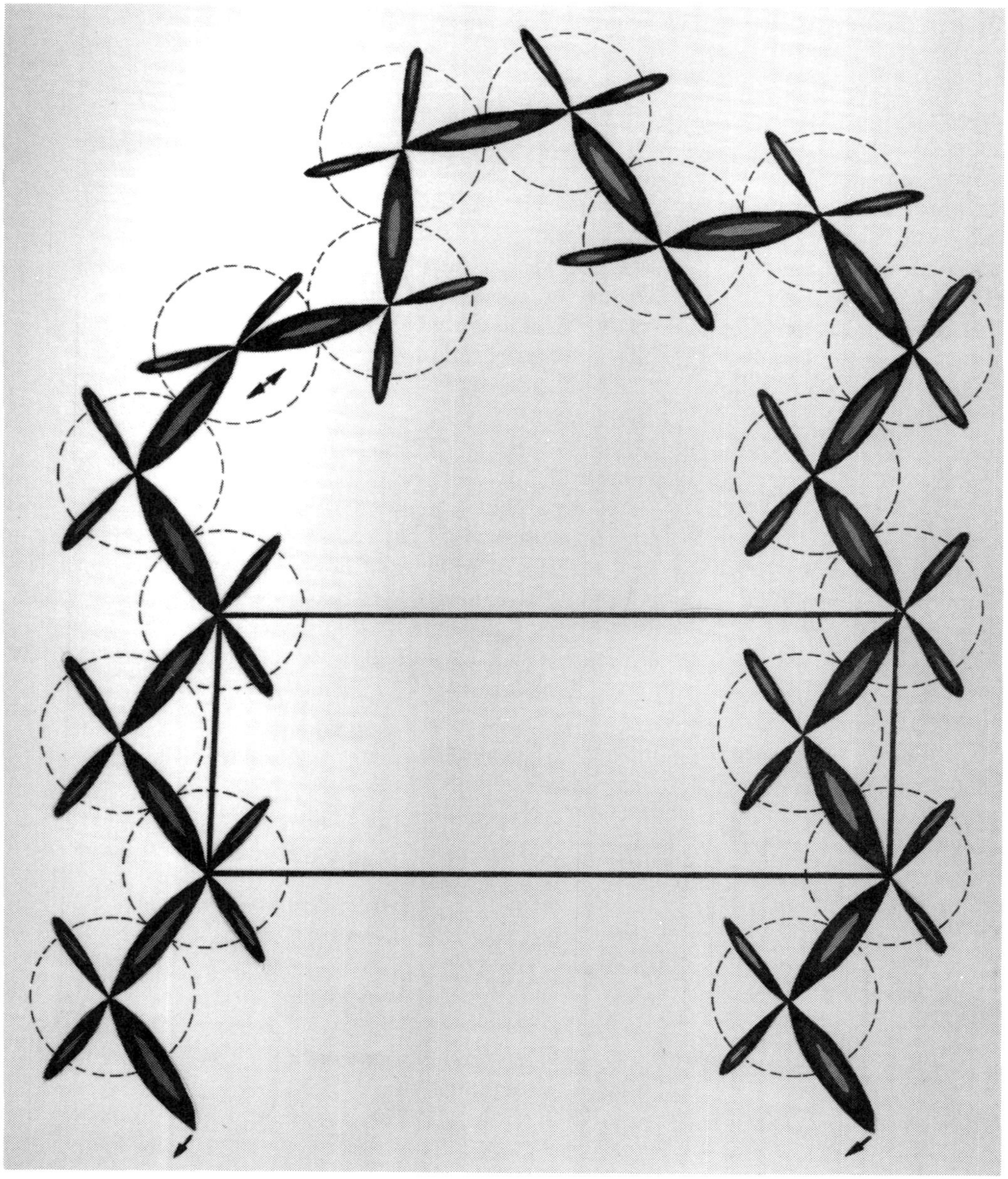

Figure 1.19. ONE FOLD IN THE POLYETHYLENE LAMELLA
This is a schematic of an individual fold of polyethylene in a lamella. A larger view of the lamella can be seen in Figure 1.18. Note that as the chain goes into the fold, there is both twisting and an extension of the normal angular bond. All of the bond variation has been concentrated into one bond for clarity. It is indicated by a double-headed arrow. Usually any bond angle variation would be spread out among several bonds. The solid straight lines forming a rectangle show another side of the crystalline unit cell noted in Figure 1.17. Besides this arrangement, there are other possibilities for rotation within polyethylene that accomplish this fold.

tional energy in place. Self-diffusion and other mobility-related phenomena are not supported in the glassy state.

RUBBER STATE

A rubber state is exhibited by most amorphous polymers. It has a temperature range above that of the glassy state and is characterized by a greater FREE VOLUME and segmental mobility than the glassy state. In the rubber state segments of the polymer may move relative to one another but entire molecules do not change with respect to their neighbors unless extremely long time periods are considered. Whenever possible, forming is done and additives are introduced into polymers while they are in this state, due to its greater deformability and diffusion rate.

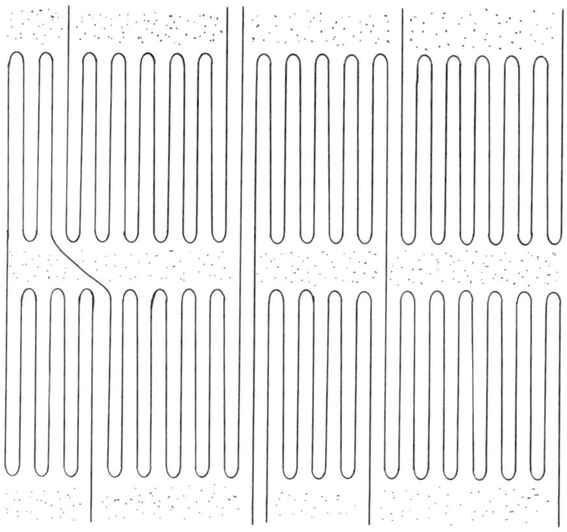

Figure 1.20. A FIBRIL OF POLYETHYLENE
This illustrates a section of a polyethylene fibril. The dotted areas are regions where amorphous polymers have formed. Along the length of the fibril are aligned lamellae, each one of which is a single crystal.

VISCO-FLUID STATE

In addition to the rubber and glassy states, there is also the possibility of the VISCO-FLUID STATE for an amorphous polymer. The visco-fluid state is just above the rubber state in temperature. It does not have the fluidity of a TRUE LIQUID, but the molecules do possess sufficient thermal mobility to move completely past one another in a relatively short time, as well as show increased segmental mobility versus the rubber state. Many polymers decompose before a sufficiently high temperature is reached for this level of mobility to be attained. This is particularly true for rigid chains with bulky pendant groups, which would resist such free motion.

GLASS TRANSITION TEMPERATURE

As an amorphous polymer goes from the glassy state to the rubber state, or vice versa, it is said to pass through the glass transition temperature. In practice, this is not a specific transition temperature but a range of temperatures over which segment mobility changes. This range is of the order of 20°C for a usual distribution of polymeric molecular weights. Because many properties change in or around this region, some of them discontinuously, it is of great importance to be aware of the glass transition temperature of a polymer.

The glass transition temperature has many attributes and all of them provide a means for its measurement. It is partly this wide array of measurement techniques

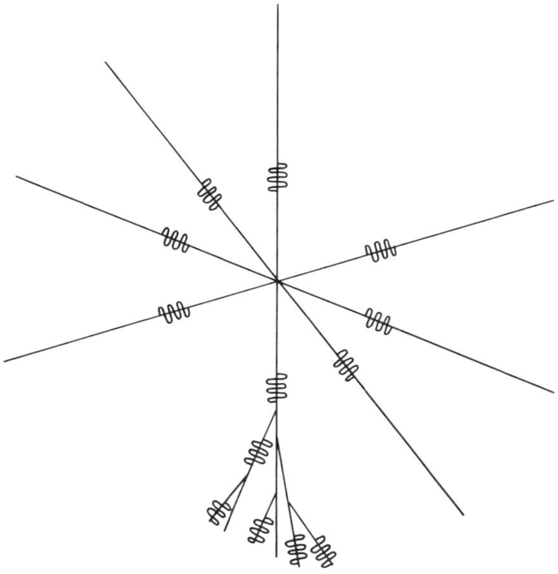

Figure 1.21. DIAGRAM OF THE SPHERULITE STRUCTURE
This is a diagram of a spherulite structure of a polymeric crystal grown from a melt of polymer. It is not characteristic of a precipitated polymeric crystal. The straight lines in the diagram indicate some of the fibrils radiating from the center of the spherulite. Along each fibril the individual lamellae are aligned in the direction of growth. When the fibril branches, as shown in the bottom right fibril, the branches are aligned at an angle with the original fibril. The growth of fibrils radiating out in all directions from the center of the spherulite, plus the branches of the fibrils, fills most of the spherical space that the spherulite will occupy if unencumbered.

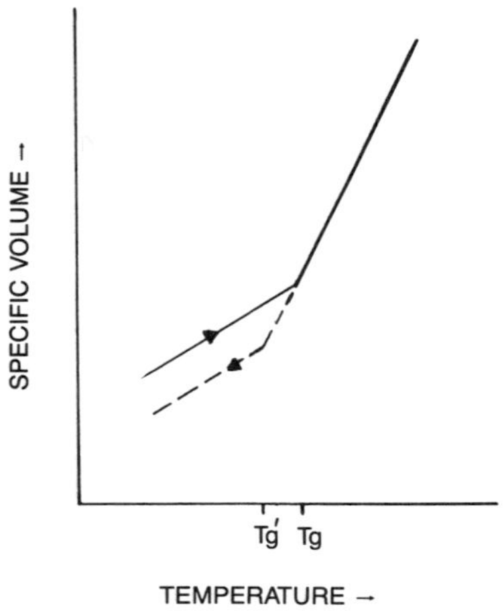

Figure 1.22. THE GLASS TRANSITION TEMPERATURE
As the temperature is decreased in a rubbery polymer, the specific volume decreases linearly until the glass transition temperature is reached. At the glass transition temperature, the rate of change of the specific volume changes as the polymer becomes glassy. The point at which the specific volume changes slope, or the glass transition temperature, is at a higher temperature if the polymer is cooled from the rubber state than if the polymer is heated from the glassy state. The former is marked as Tg and the latter is marked Tg′.

and partly the fact that the glass transition temperature is really a range of temperatures for commercial polymers that cause such wide variation in the reported values. The most commonly used method of determination uses specific volume. In this technique specific volume is measured as a function of temperature and the results are plotted. Both in the glassy and the rubber regions the specific volume varies linearly with temperature, but the slopes of the two states are not equal, the rubber state always having a greater slope than the glassy. The intersection of the slopes is declared the glass transition temperature. Figure 1.22 shows a graph of this. A similar procedure can be followed to determine the FLOW TEMPERATURE, or the temperature at which a polymer goes from the rubber to the visco-fluid state.

FACTORS THAT INFLUENCE THE GLASS TRANSITION TEMPERATURE

Once the molecular weight of a polymer is sufficiently large that the material can be used for structural purposes, the molecular weight and its distribution are not of great importance to most macroscopic properties. The glass transition temperature is one property that is sensitive to molecular weight and the distribution of that weight. As the molecular weight of a polymer increases, so does the glass transition temperature of that polymer. This follows, since the larger the molecule the more difficult mobility becomes and consequently the greater the temperature required to give the polymer sufficient energy to move in a way appropriate to the rubber state. An example of making a molecule much larger, and thus appreciably raising its glass transition temperature, can be seen when a polymer is lightly cross-linked to effectively produce huge molecules of the same chemical composition as the original polymer. In this case the glass transition temperature increases linearly with cross-linking. Alternately, when lightweight polymers are introduced, usually by processing, the effective glass transition temperature is lowered by a calculable amount. Thus, it is the molecular weight of the polymer that determines the value of the glass transition temperature and it is the variation in molecular weight within a polymer that most directly produces a range for the glass transition temperature. An extreme example of raising the glass transition temperature would be a highly cross-linked polymer that remains glassy until decomposition. If this highly cross-linked polymer were then saturated with a lightweight polymer, with its correspondingly low glass transition temperature, the result would be a polymer with a glass transition temperature range of the hundreds of degrees that would normally separate these two different polymers.

One of the characteristics of the glass transition temperature is that the free volume fraction is 0.025. Any factor that influences the free volume of a polymer will thus influence the glass transition temperature. Accordingly, when a polymer is put under pressure the glass transition temperature will increase with increasing pressure, usually linearly. Polymers with large pendant groups often have variations in density depending on the regularity of the pendant group arrangement. If the pendants are randomly arrayed, the density is lower than in a regular array, and the corresponding glass transition temperature will be lower for the lower density material. However, when the regularity of the structure does not produce a significant difference in the density, a greater variation than about 2%, then the glass transition temperatures of the ordered and non-ordered structures are, about the same. This would make density variation a good variable to consider when estimating expected glass transition temperature of a polymer if there are no additional structural or bonding considerations.

For crystalline polymers, the crystal segment does not undergo a glass transition temperature, but the surrounding amorphous regions do. For this reason, a glass transition temperature may be listed for a mainly crystalline polymer. In most crystalline polymers the density of the crystalline segment is greater than the amorphous region. Thus, crystallinity in a polymer tends to lower the free volume of the entire polymer and in turn raise the glass transition temperature of the amorphous region. The increase in glass transition temperature due to the presence of crystals in the polymer is usually a linear function.

Knowledge of the structure of a polymer may help to anticipate its glass transition temperature, but there are no useful formulas which can be easily applied. Factors like stiffness of the backbone, bulkiness of the pendants, branching, cross-linking, symmetry, and polarity all affect the glass transition temperature. Of these factors, stiffness of the backbone is one property which would not be readily obvious. The others will allow at least qualitative comparisons between polymers. Comparing a polymer to one of known glass transition temperature can provide an expected trend if not an actual value.

ADDITIONAL TRANSFORMATIONS

Below the glass transition temperature may exist additional transformations that are secondary to the main transformation. As an example, in the rubber state a segment of the chain that rotates usually involves about twenty chain atoms but when the polymer is cooled below the glass transformation temperature this number drops to three or four atoms in the segment which move together. Further cooling of the polymer below the glass transformation temperature would produce a secondary transformation below which all segmental motion would cease.

Above the glass transformation temperature there are two additional transformations possible which are of the same order as the glass transformation. These are: the visco-fluid transformation when a rubber becomes visco-fluid and the liquid transformation when a visco-fluid becomes a true liquid. These transformations, especially that of the liquid transformation, have not been well characterized, since they are much more structural than thermodynamic in nature and not all polymers display a visco-fluid and a liquid transformation. What is usually reported is the DECOMPOSITION temperature since it is both thermodynamic and structural in nature. Here there is no difficulty in defining the resulting gaseous state. Sublimation does occur slowly under all conditions but is greatly accelerated when the polymer reaches elevated temperature. The rate of sublimation can also be increased by chemical attack, reduced pressure, or any other factor which causes the breaking of chemical bonds. Therefore, the sublimation temperature, like the glass transition temperature, is really a range of temperatures and can vary widely with conditions. Even once a polymer is said to sublime, additional breakdown, which can be considered transformation, is possible. On sublimation, polymers usually break up into small molecules rather than go directly to atomistic form. The breaking up of these molecules into atoms can occur at about the same temperature, causing one more transformation.

The crystalline part of a polymer does exhibit a melting point, although it is rarely as precise as melting points in other materials. When the polymeric crystal is heated toward the melting point the smaller and less perfectly formed crystals reach a point of liquidity before the larger or more perfectly formed areas do. As the former begin to loose stiffness, the polymer shows both structural and thermodynamic evidence of localized melting. Thus, the reported melting point of a polymer is usually a median temperature within the range of temperatures over which the polymer actually melts. Accordingly, annealing a crystal to promote crystal growth and reduced defects will increase the melting point of the crystal, whereas introducing branching and other imperfections will decrease the melting point of the crystal.

Besides the set of transformations that exists for amorphous polymers and the set that exists for the crystalline segment of polymers, it is also possible to transform some amorphous polymers into crystalline ones without melting them. This can be achieved by stretching the polymer and thus bringing into alignment those segments of the polymer chains that have previously been too far from alignment to crystallize. It is only practical from the mechanical properties and relaxation time point of view to do this above the glass transformation temperature.

TRANSPORT IN POLYMERS

Since the glass transition temperature corresponds to a mobility change in a polymer and has a definite free volume associated with it, it is not surprising that the JUMP FREQUENCY of a segment at the glass transition temperature is the same magnitude for nearly all polymers, about 0.1 per second. The diffusion rate of the segments increases approximately linearly with increasing temperature except at the transition temperature where the rate of change alters. Below the glass transition temperature the DIFFUSION RATE of the polymers segment is too low to be useful within most time frames. Above the glass transition temperature the SELF-DIFFUSION rate continues to increase such that the useful annealing temperature is well below the next transformation temperature.

Inter-diffusion with another material is dependent on the characteristics of both materials. The most common form of inter-diffusion is between a polymer and a lighter weight organic liquid, as in the use of solvents for processing. Above the glass transition temperature the polymer has greater free volume and more mobility. Thus, in this temperature range there is more space for the diffusing material to move into and a more likely path with the segments moving.

Very small molecules of either liquid or gas passing through a polymer are usually considered to penetrate it rather than diffuse through it. This penetrating ability or PERMEABILITY is highly dependent on the size of the penetrating molecule. Very small gases can pass quickly through the polymer since the polymer contains many open spaces of a small size. As the size of the penetrating molecule increases, the number of sites available to move into decreases rapidly and the penetrating molecule must await the opening of such a site by segment motion of the host polymer. Thus, the time required for a suitable site increases in a non-linear fashion with the increase in the size of the penetrating molecule. Because of this, polymers have been used as molecular filters.

SOLUBILITY

As an almost universal rule, linear organic polymers have a SOLVENT. This is true for branched polymers as well, since the branch sites allow for additional free spaces for the solvent to move into. However, the amount of absorption of a solvent may not be as high in a branched polymer as in a linear polymer. Cross-linked polymers, which are SOLUBLE since they are based on linear polymers, have an even more limited absorption ability called SWELLING. With increasing cross-linkage the swelling decreases due to reduced free areas for the solvent to move into. Similarly, other factors that tend to reduce free area in a polymer, like crystallinity and increased pressure, will reduce the amount of solvent absorbed. At the other extreme, the increased surface area of a foam and the internal stresses associated with formation make foams more soluble than their basis polymer.

DISRUPTION OF THE CHEMICAL BOND

The factors that break chemical bonds in other materials also serve to break bonds within polymers. However, because these bonds are optimized throughout the polymer, the conditions necessary to break the bonds must be more severe than they would be to break the bonds within a smaller molecule. When conditions are favorable for bond breakage within a polymer, it is more catastrophic since many bonds will be of similar strength and break at the same time. If the bonds within the pendant groups of a polymer are the bonds affected by the bond-breaking process, then the overall properties of the polymer will not be greatly affected. Properties are much more sensitive to bond breakage along the main backbone or inter-molecular cross-linking.

Organic polymers are particularly sensitive to THERMAL DEGRADATION. Their service temperature is usually below 475° Kelvin and they almost always decompose below 1300° Kelvin. A useful application of this limited temperature tolerance is found in the aerospace industry. THERMOSETTING polymers are used as sacrificial coatings since they ABLATE, forming a surface with a very low heat transfer rate.

Chemical bonds may also be broken by agents which cause a chemical reaction. These agents, or reactants must reach the bond sites in order to cause the disrupting bond breakage so physical proximity is usually the limiting factor in this type of degradation. If the reactant is soluble in the polymer then it can diffuse throughout the polymer and react at all favorable sites. Thus, factors that decrease solubility and diffusivity of the reactant decrease the damage to chemical bonds of the polymer. Two common reactants found in the air are oxygen and water vapor. Both are usually soluble in polymers and small enough to diffuse freely. The reactions they produce with the polymer are slow due to the necessity of breaking existing bonds first. Although they do limit the service life of a polymer, they are not considered severe in themselves. It is only when an energy source to break existing bonds is added that the reactions become much more rapid. A prime source of energy to encourage the reaction of polymers with oxygen, water vapor and other CORROSIVE reactants is sunlight. Polymeric molecules are particularly susceptible to bond breakage by ultraviolet light in the 300 to 400 nm wavelength range. Once the bonds are broken, the radicals are free to react with oxygen, etc. Fortunately, this wavelength range only amounts to 5% of the sunlight on the earth's surface. Also, the penetrating power of the ultraviolet light is limited, so the destruction is mainly a surface phenomenon.

PARTICLE RADIATION, or ionizing radiation, has a much greater penetrating ability in polymers than ultraviolet light. Thus, its effects will be felt throughout the bulk of the polymer rather than just near its surface. This characteristic can be put to positive use in polymers in which the bond breakage encourages spontaneous reactions resulting in cross-linking of the polymer chains. Additionally, particle radiation reacts with individual atoms, unlike ultraviolet radiation which affects the entire molecule.

Additives 2

ADDITIVES

ADDITIVES include all substances that are intentionally added to a polymer mixture, but which do not form primary chemical bonds. Like alloying elements in metals, additives are introduced to alter properties. However, volatile additives may also be included during processing and may not be in evidence in the final product. Many volatile processing additives do leave some residue, which can affect impurity-sensitive characteristics such as light absorption. It is worth remembering that anything put into a polymer mix is likely to have an effect, although the specific effect and its magnitude can be difficult to predict.

Some additives change the entire characteristic of a polymer by their presence. An example of this is the mixing of a red colorant into a clear polymer resulting in a red product. Other additives appear to modify the entire material when only the surface has been affected. This can be exemplified by the addition of the red colorant in the previous example to only the surface of the clear polymer. The product still appears to be entirely red but much less additive has been used. The third most common case is the addition of an additive to only one surface. Using the red colorant in the clear polymer example again, the red colorant would only be applied to one surface of the polymer, so that it would appear red only from above or below the surface to which the colorant was added but not from the other sides. Such an approach further conserves additive.

Solids, liquids and gases can be used as additives. Solids are usually added to the polymer when it is in its most fluid state and are mechanically mixed into it. One of the problems associated with solid additives is clustering of the solid rather than uniform dispersion throughout the polymer. Liquid additives generally are of small size relative to the polymeric chains and chemically similar. If there is a sufficiently large size difference between the polymer and the liquid additive, the additive will only be able to dissolve into the polymer a segment of the chain at a time. It could then be a slow process for the liquid additive to dissolve the polymer segment by segment. Gases often become additives due to their generation during the processing of a polymer into its final form. If it is considered desirable to retain these gases or introduce other gases, the difficulty lies in producing a polymeric structure which can contain them, since most structures allow gases to diffuse through them readily. This is a reoccurring problem in the production of polymeric foams.

PLASTICIZERS

PLASTICIZERS are additives used to increase the FLEXIBILITY or plasticity of the polymer. Occasionally they are used only to facilitate the processing of a polymer and are not meant to permanently change the flexibility of the polymer. Then a VOLATILE plasticizer is used so that it will evaporate once the final product is produced. The smell associated with new plastic is often caused by such volatile plasticizers.

The mechanism by which plasticizers work is to intersperse themselves around the polymer and interfere with the chain-chain secondary bonding. This interference reduces the inter-chain bonding strength and allows the chain segments greater mobility resulting in increased flexibility and decreased TENSILE STRENGTH. Usually a plasticizer is chosen for its solubility in the polymer, since a soluble plasticizer can penetrate the polymer more easily than a non-soluble one. Plasticizers can also be of the non-soluble type and merely mechanically dispersed throughout the polymer. This type of plasticizer is not as effective at increasing flexibility for its weight fraction and has the increased disadvantage of extensive EXUDATION at the surface.

Most liquid plasticizers are low molecular weight organic materials with a glass transition temperature in the range of 125° to 225°K. When this type of solvent is added to an organic polymer the result is a weight averaging of the glass transition temperature between that of the polymer and that of the plasticizer if they are miscible. The reduction of the glass transition temperature of the polymer by the plasticizer/solvent under the above conditions is usually about 5° Kelvin per weight percent of plasticizer.

The discussion given in Chapter 1 relating free volume to glass transition temperature can be applied to plasticizers. The small molecules of the plasticizer open up the structure of the polymer thus increasing its free volume at all temperatures. Since the free volume required for the glass transition to occur is a constant, the transition would be reached at a lower temperature for a plasticized polymer.

Additives of any type will affect the overall properties of the polymer. However, with plasticizers the only desirable effect is usually an increase in flexibility. For this reason, plasticizers should only be used as needed and in the minimum amount. Plasticizers are subject to AGING, EXTRACTION, and MIGRATION as well as volatility and exudation; all of which limit effectiveness and still leave traces of the plasticizer that can be detrimental to properties. In addition, most plasticizers are toxic and only a very few are approved by the Federal Food and Drug Administration for use with foods and medicines.

FILLERS

The term *filler* has previously been used to define virtually anything added to a polymer for purposes other than plasticization. More recently, the term has come to mean a non-reacting additive used primarily to add bulk. As such, the ideal filler would be cheap, pure, non-flammable, white, soft and of uniformly small size, so as to minimize its possible degradation of a polymer's properties. Calcium carbonate is a widely used filler that fits these criteria. Other minerals are also used, but minerals in general have the problems of AGGLOMERATION and irregular hard surfaces that cause abrasiveness. Organic materials, like wood flour and coconut-shell flour, are often used as fillers. These materials give some reinforcement to the polymer and therefore should be considered reinforcers. Any filler can become a reinforcer under hydrostatic pressure, since such pressure forces an increase in the van der Waals bond between the filler and the polymer, with a corresponding reduction in free space.

REINFORCEMENT ADDITIVES

Powders and fibers added to a polymer may serve to reinforce or increase the tensile strength of the polymer. Most REINFORCEMENT ADDITIVES are bound to the polymer by van der Waals forces and, thus, are better bonded if the surface of the particle is smooth and the surface area and surface to volume ratio are high. This, then, would be from small, spherical particles.

Small particles are further favored if one considers stress concentrations within the polymer. Large irregularly shaped particles can initiate tearing when a force is applied to the polymer. Also, small particles more uniformly distribute the reinforcement for the same weight of additive and reduce the FRESNELL EFFECT. Thus, small particle size is always favored over large particle size for reinforcing additives when the particles are uniformly distributed. If, however, the small particles agglomerate, they are tantamount to large irregularly shaped particles and can actually weaken the polymer. See Figure 2.1 for an illustration and discussion of particle shape.

Other factors that affect the tensile strength of the reinforced polymer are the bonding strength between the polymer and the particle, and the relative strengths of the polymer versus the polymer–particle bond. If the polymer–particle bond is weak relative to the tensile strength of the polymer, then the polymer will pull away from the particle at a LOW STRESS, resulting in STRESS WHITENING and limited reinforcement. When the polymer–particle bond strength is about that of the polymer–polymer bond the particles tend to flow with

Figure 2.1. THE SHAPE OF ADDITIVES
The four shapes to the left occupy the same volume but have markedly different effects when used as additives. The first shape on the left has the problems of both sharp corners and an irregular shape, which is ineffectively surrounded by polymer chains. The second shape from the left is better since it does not have sharp points that might sever chains. The rod shape of the third can be very effective when there is polymeric orientation along the rod length. However, this shape is not advantageous in some applications where three-dimensional reinforcement is to be optimized, since it has a large surface area perpendicular to the axial direction and weakens the polymer-polymer bonding. The fourth and last shape, the sphere, is the most universally useful shape.

Figure 2.2. ENCAPSULATION OF AN ADDITIVE PARTICLE
The diagram on the left shows a spherical particle surrounded by parts of five different polymeric chains. In this case the polymeric chains do not tightly enclose the particle. The effect of shrink fitting can be seen by the illustration on the right which is the same except for the closer proximity of the additive to the polymer.

the polymer during deformation, thereby increasing the overall tensile strength of the material without altering the yield point. However, the ultimate elongation will be increased since the strength is higher, but the particles do not interfere with the viscous flow of the now stronger material. In the rate case of the polymer–particle bond strength being higher than the strength of the polymer–polymer bond, the polymer is restrained from moving around the particle resulting in a stronger, more rigid material.

A tailoring of the polymer–particle bond strength to the particular polymer is clearly desirable. In the usual case, the polymer is organic and the reinforcing particle is a ceramic or metal with nothing but very weak mass attraction for inter–bonding. Furthermore, the polymer may be stiff enough to prohibit flow around the particle and encapsulate it tightly. Obviously, the closer the polymer and particle, the stronger the mass attraction. The encapsulation problem can be helped by adding the ceramic or metal particles at a temperature higher than the service temperature and allowing the polymer to shrink around the particles, as the rate of thermal contraction is universally higher for organic polymers than ceramics or metals. Figure 2.2 illustrates encapsulation of a particle.

The polymer–particle bond can be further improved by the use of COUPLING AGENTS. These chemicals form a bond with the polymer and the particle, which serves as a bridge between the two. A further refinement of this idea is the CERAPLAST. Here the normal coupling agent thin film is increased to a relatively thick layer of tough rubber-like material. The bond between the layer of the ceraplast and the ceramic reinforcement particle is primarily mechanical, but, since there is no requirement for flow, this proves to be more satisfactory than direct encapsulation of the particle by the polymer. The thick encapsulating rubber not only bonds easily to the polymer, but also reduces stress concentrations associated with particle reinforcement. Figure 2.3 illustrates ceraplasts in a polymer.

ELASTOMER REINFORCEMENT

Elastomers that are not crystallizable are often reinforced to enhance their mechanical properties. This is usually done in conjunction with cross-linking. The simple addition of an inorganic reinforcement agent, however, does not have as favorable an effect on the

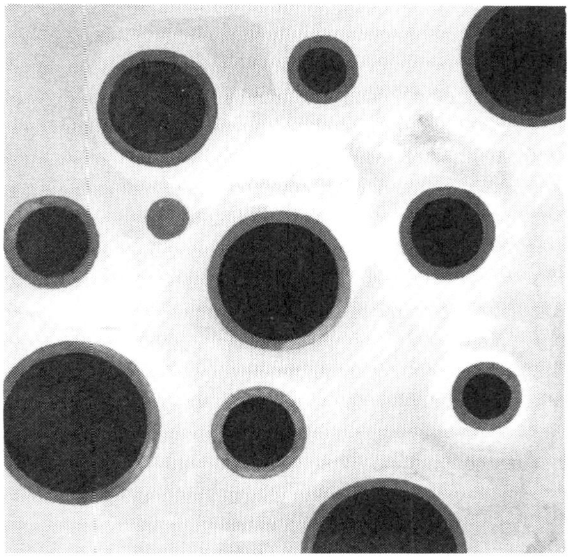

Figure 2.3. CERAPLASTS IN A POLYMER
This illustration shows how ceraplasts in a polymer look if a slice through the bulk is taken. The pale gray background depicts the polymer and the gray represents the rubber-type coating on the ceramic reinforcement additive, shown as the black circles.

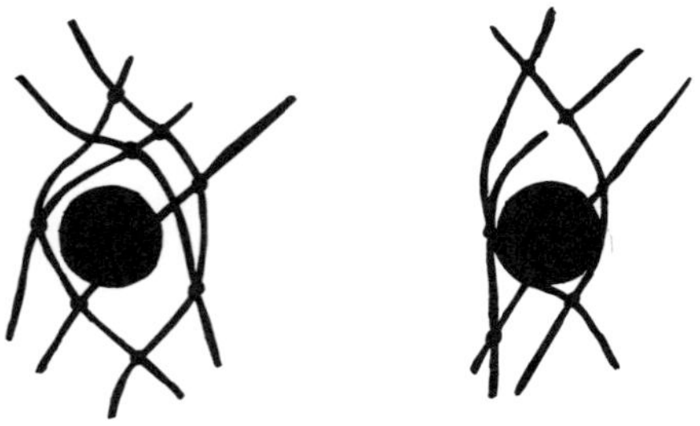

Figure 2.4. REINFORCING ADDITIVES IN ELASTOMERS
In this diagram the black circles are the additives and the lines are the polymers with cross-linking sites marked with dots. The diagram on the left shows the additive moving with the cross-linked elastomer. The additive on the right does not move with the matrix and causes stress within the chains, such that breakage can occur. One such broken chain is included.

properties of elastomers as it has for other polymers. Elastomers require a reinforcing additive that has a higher bonding strength than that available through just mass attraction.

Two additives that form a stronger secondary bond than just mass attraction are silica and CARBON-BLACK. Silica bonds well with silicone elastomers due to their chemical similarity, while carbon-black is an effective reinforcing additive for nearly all carbon-based elastomers for the same reason. The mechanism for silica is not clear but carbon-black bonding seems to be due to a weak covalent bond.

As would be expected, the strength of the reinforced elastomer goes up with loading. However, if the dispersion is poor, which is often the case with silica, then there is no simple relationship between strength and loading. Strength improves with loading for well-dispersed and bonded additives because they then have sufficient bonding strength to move with the polymer they reinforce. In the case of high agglomeration or weak bonding of the additive, the strength can remain unchanged with increasing loading, or the strength may decrease due to stress build-up and local chain breakage. See Figure 2.4 for an illustration of this. Although these additives may give the desirable result of increased strength and abrasion resistance, they can also produce stress-softening (the MULLINS EFFECT), compression set, hysteresis and heat build-up. These negative effects are also in proportion to the loading.

Elastomers that crystallize on stretching, like natural rubber, show little improvement, no improvement, or deterioration, when a reinforcing agent is added. This is probably due to the nature of the reinforcement. When the crystallizable elastomer is stretched, it improves alignment of the chain segments resulting in crystallite formation. These crystallites then act as hard regions in the pliable amorphous polymer and serve to reinforce it. Additives disrupt the polymeric order normally present on stretching an elastomer and reduce the amount of crystallization possible. Thus, although a reinforcing additive may continue to reinforce the elastomer, the reduction in crystallites, which the elastomer suffers due to the presence of the additive, tends to offset the effect. Additives in crystallizable elastomers continue to display all of the negative attributes. As a consequence, they are very rarely used in crystallizable elastomers.

SURFACE PROPERTIES MODIFIERS

Nearly all additives alter surface properties. A very common modification is the production of surface roughness. This comes about if the thermal contraction of the polymer differs from that of the additive. The polymer is almost always processed at elevated temperatures. Thus, a differential in thermal contraction results in roughness. The roughness can, however, be an advantage if less surface gloss and adherence to other polymeric surfaces is preferred. In addition, a reduction in surface adherence due to roughness also reduces the friction between adjacent polymeric parts. See Figure 2.5 for an illustration.

The amount of roughness due to additives can be controlled by the amount of differential thermal contraction between the materials, as well as the size,

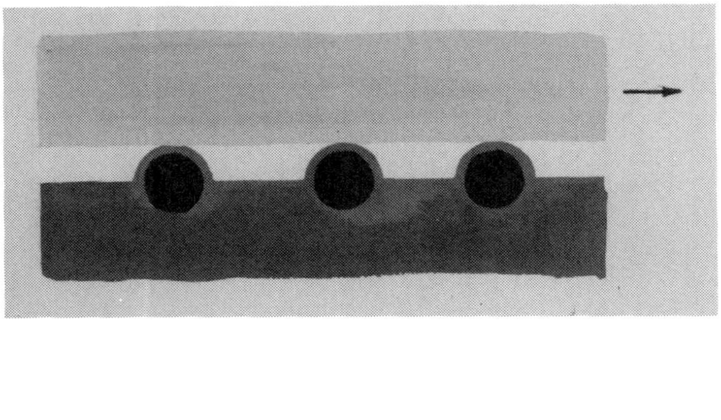

Figure 2.5. FRICTION BETWEEN POLYMERIC SURFACES
In this exaggerated view of the surface of two polymeric materials, the upper material, light gray, is being moved across the bottom material, dark gray with black additives, in the direction of the arrow. Note that the surface contact between the upper and lower polymers is reduced, due to the presence of additives in the bottom polymer. As a result, the coefficient of friction and adherence between these two surfaces are reduced.

Figure 2.6. ABRASION OF POLYMER BY ADDITIVES
As in Figure 2.5, this is an exaggerated view of the surfaces of two polymeric materials. The upper material, light gray, is being moved across the bottom material, dark gray with black additives, in the direction of the arrow. In this case, the surface contact is also limited but the coefficient of friction and adherence are both high because the jagged edges of the exposed additives are cutting into the surface of the upper polymer making motion difficult. This increases the coefficient of friction and the adhesion between the surfaces and abrades the upper surface.

shape, hardness, and distribution of the additive in the polymer. Small differences in thermal contraction produce slight roughness on the surface. Large, hard, irregularly shaped additives are the most conducive to surface roughness and thus have the lowest adherence and friction. However, if an additive cuts through the surface of the polymer and is exposed, then the exposed part of the additive can act as a cutting agent and both abrade other surfaces and increase the coefficient of friction of the surfaces. See Figure 2.6 for an illustration.

Polymer surfaces are particularly subject to static charge build-up since a static charge that is produced cannot be easily dissipated due to the very poor electrical conductivity of most polymers. The build-up can be minimized through reducing the surface contact between polymers, e.g., by increasing surface roughness. Also, the use of additives such as carbon-black to improve electrical conduction will aid in static charge dissipation. Carbon-black has also found use in polymers in reducing friction, since it acts as a solid lubricant when exposed.

The loss of transparency due to scratching is a serious problem with all optical grade polymers. None of these have a hardness in excess of 3 mohs, whereas most dust has a hardness of about 5 mohs. Thus, dust and most other common particles are highly abrasive to polymers. Hard and transparent coatings like glass and quartz have been used in small patches to protect the surface of many polymers. Extensive use of these coatings has not been made due to the great difference in the coefficient of thermal expansion between the coating ceramic and the underlying polymer, resulting in the flaking of the coating during temperature changes.

OXYGEN STABILIZERS

Atmospheric oxygen is a primary potential reactant for any polymer. However, the actual reaction rate of oxygen with a polymer depends on many factors and is usually quite sluggish. One of the factors that influences the reaction rate is the inherent chemical stability of a polymer, since a polymer must first break existing bonds in order to form new ones with oxygen. If a polymer is loosely bonded and has a correspondingly low melting point, then it can be expected to react more readily than one more tightly bonded. Elevating the temperature of the polymer gives it more energy and allows for easier bond breakage, whereas lowering the temperature has the reverse effect. Exposure to energy sources such as light increases the likelihood of bond breakage. Additional factors that affect reaction rate are the presence of catalysts, such as some metals, and the diffusivity of the oxygen through the polymer, since it must be in contact with the reaction site in order to react.

When oxygen does react with a polymer the first stage is almost always chain scission. Following this, the chain may remain in smaller segments or it may re-react to form a cross-linked structure. When only chain scission occurs, the average chain length is greatly reduced, causing the polymer to display the characteristics of a lighter, weaker structure. Conversely, if cross-linking is the final stage, then the resulting polymer is larger, stronger and more brittle. Regardless of the specific results of the oxidation, it will ultimately be detrimental to the long-term applicability of the polymer, if it continues to oxidize in an uncontrolled manner. ANTIOXIDANTS are added to particularly susceptible polymers in an effort to control the oxidation. These are usually organic additives that absorb either the oxygen itself or the energy necessary for the reaction with oxygen to occur. Additional discussion of the oxygen-polymer problem will be given in the following three sections.

ULTRAVIOLET LIGHT STABILIZERS

The shorter wavelengths or ultraviolet portion of the sunlight spectrum are able to break most covalent bonds within a polymer. The amount of bond breakage depends both on the exposure time to sunlight and on altitude. Extra-terresterial exposure is particularly harmful since there is no cloud cover or atmosphere to absorb the harmful wavelengths. Once the bonds are broken, the polymer is subject to reforming in an uncontrolled manner and reaction with oxygen and other atmospheric gases.

Ultraviolet light stabilizers can be metallic materials which preferentially absorb or reflect the ultraviolet light. Such metallic materials are opaque and transfer to the polymer their characteristic color. A commonly used and very effective additive which is loosely in this category is carbon-black. When color is of concern, organic materials can be used as stabilizers. The organic materials most useful as ultraviolet light stabilizers have the ability to absorb the harmful wavelengths and re-emit the absorbed energy with a wavelength that will not cause bond breakage. To be effective, the organic additives must be on the surface of the polymer as well as mixed through it. Many of the organic ultraviolet light stabilizers, such as 2-hydroxybenzophenone, are toxic, and this limits their use.

HEAT STABILIZERS

Because the primary reason for failure of a polymer at elevated temperatures is chemical decomposition, which is usually a direct reaction with atmospheric oxygen, heat stabilizers are often considered to be oxygen stabilizers. However, a heat stabilizer may also be used to control the temperature of the polymer so that structural changes do not occur. For example, color additives serve to trap specific wavelengths of light which can lead to thermal build-up in a nonconductive polymer. If this happens the polymer may react by shifting double bonds locally or by changing angular relationships within the chain. Such changes will then affect the properties of the polymer. The primary means of producing a more thermally stable polymer is the addition of a heat stabilizer that will allow for the energy that is trapped in the polymer to be dissipated before the temperature is raised to too high a level. This is most commonly done by adding metal powders.

FLAME RETARDANTS

It is possible for the reaction rate of a polymer with the oxygen in the air to be so great as to produce a flame, while fragments of the polymer come off as smoke. This usually occurs only at high temperature and once the polymer is partially broken down. The addition of flame retardants can reduce this reaction with oxygen below the flash point, but it cannot eliminate it.

One obvious way to reduce the risk of flame is to remove the polymer from the presence of oxygen. This can be done either through a protective environment or a coating. Any coating which will prevent the diffusion of oxygen through it and meet the other requirements of the design will suffice. However, the most effective materials at preventing oxygen diffusion are metals or ceramics of a thickness on the order of 1 millimeter or more. This type of coating may not be easily incorporated into the design or easily cladded to the polymer.

The chemical composition and structure of the polymer determine its potential flammability, both by the amount of energy required to break the internal bonding and by the type of product that results. Polymers that produce combustible gases such as carbon monoxide have a volatile product that can further react in the vicinity of the flame and cause the flame temperature to increase. Polymers that produce gaseous products such as hydrogen chloride, which are non-reacting, tend to have higher flash temperatures, since additional heat is not being supplied by this type of product. In addition, the non-combustible gases help to dilute the concentration of the oxygen in the vicinity of the flame, limiting its spread. Additives that are effective at producing non-reacting products are trihydrate and magnesium hydroxide. Polymers that char on burning have a natural means of limiting flame production and are said to be self-extinguishing. The char acts as a barrier to incoming oxygen and limits escape of the outgoing combustion products. Most cross-linked polymers produce a char on burning. The presence of a halogen element in the polymer or in an additive will act as both a flame and smoke suppressant. The halogen will usually be more effective in the presence of a metal oxide. As energy is added to a halogen-containing polymer, the decomposition that occurs will produce halogen-bearing products that are dense and non-reacting. Thus, there is no noticeable smoke and no secondary oxidation to increase the energy.

Smoke suppressants are often used in conjunction with flame retardants. Most smoke suppressants work by breaking down the volatile part of the polymer into sufficiently small units that they become colorless gases. The smaller molecules escape from the polymer more readily than larger molecules and thus are removed from the possibility of further combustion.

MICROORGANISM SUSCEPTIBILITY

Polymers which contain carbon obtained from an

organic based monomer are potentially susceptible to attack by microorganisms such as fungi and bacteria. In addition to organic carbon, microorganisms need warm temperatures, moisture and protection from intense ultraviolet light. For a large number of polymeric applications, all of these conditions will obtain. However, the degree of susceptibility varies greatly. Hence, the avoidance of microorganism attack may depend primarily on the wise choice of a polymer rather than the use of an additive. Polymers that are close to their original structure, such as cellulose and natural rubber, are much more subject to attack than the more processed polymers, such as nylon. In addition, crystallinity protects a polymer from microorganism attack.

Even highly resistant polymers may become BIODEGRADABLE if they have additives that are subject to microorganic attack. The most common problem in this regard is plasticizers. Again, the most straightforward approach is to avoid the susceptible additive. When this is not possible, then fungistatic and bacteriostatic additives can be used. The most effective seem to be the organo-mercury compounds. However, there is a risk of toxicity associated with these compounds. An acceptable alternative is usually a chlorinated compound.

COLORANTS

In order to produce color in normally clear polymers, DYES, and pigments are added. Dyes are usually soluble, fine-sized, organic particles which produce brilliant transparent color that is much more susceptible to degradation by chemicals and light than pigments. Pigments are much more capable of hiding flaws than are dyes since they are mainly metallic oxides and opaque, but pigments have much lower coloration capacity for their weight than dyes. Pigments often have undesirable effects on the mechanical properties of the polymer and cause processing difficulties due to their large size and non-bonding nature. As a result, pigment concentration is kept below five percent.

Physical Properties 3

DENSITY

Polymers as a group are among the least dense materials, with specific gravities ranging from about 0.9 to 2.0. Foams have even lower effective densities, while polymers with dense additives, such as metal-oxide dyes, may exceed the range given. The density of most polymers is about 1.2 grams per cubic centimeter, or less than 20% of the density of steel. Such a low density makes polymers easier to handle, cheaper to transport, and highly advantageous in designs that have weight limitations. Closed cell foams are particularly useful in this respect because their continuous surface prevents the absorption of liquids, making them applicable for floatation.

THERMAL EXPANSION

Polymers are generally considered to have a nearly constant THERMAL EXPANSION rate as a function of temperature over their entire usable temperature range. This allows a single value of expansion rate to be a useful approximation for most calculations. Nevertheless, the coefficient of thermal expansion (CTE) is always less for the glassy phase than it is for the rubbery phase, and there is a similar difference between the rubbery and the visco-fluid phases. The crystalline phase displays a constant coefficient below the melting point, that is about the same value as the CTE of the glassy phase for that material. Although the values for the coefficient of expansion for different types of polymers vary by as much as an order of magnitude, the difference between the coefficient for the glassy phase and the rubbery phase at the glass transition temperature of any particular polymer is usually a consistent value for all polymers. This fact is used in support of the free volume theory of glass transition temperature.

The coefficient of linear thermal expansion for polymers is around five times that of metals and around ten times that of ceramics. The large difference in thermal expansion between these materials often poses problems in design when dissimilar materials are to be joined. Warping and joint separation are common for polymers under these conditions. One thing that can help to alleviate the problem is the presence of additives in the polymer. Inorganic additives will reduce the coefficient of thermal expansion of the polymer to which they are added and bring the resulting material into better agreement with the other materials to which it is to be joined. Alternatively, additives of organic materials, especially low molecular weight plasticizers, serve to increase the coefficient of thermal expansion of the polymer to which they are added.

HEAT CAPACITY

Most polymers have a SPECIFIC HEAT of about 0.4 calories per gram and a density of 1.2 gm/cc which means that the HEAT CAPACITY by volume would be about 0.5 cal/cc. Although this value of heat capacity has been calculated for a pure polymer, it is applicable to a polymer with inorganic additives too. In the case of inorganics, the specific heat is lower but the density is higher, so the resulting heat capacity tends to be about the same value as that of a polymer. Carbon-black is an exception to the normal value of heat capacity for an inorganic material with a value of about 0.2 cal/cc. In a polymer-carbon-black combination, the resulting heat capacity can be estimated by considering the relative contributions of the materials on a volumetric basis. Similarly, metallic additives, which usually have very high heat capacities, can be considered on a volumetric basis for their contribution to the heat capacity of the final material. In the case of a gas additive the heat capacity of a gas is so low that it is negligible for estimating purposes. Thus, the heat capacity of a foam will be the heat capacity of an equivalent volume of the polymer rather than the actual volume that the foam exhibits.

THERMAL CONDUCTIVITY

Thermal energy, which is primarily vibratory in the solid state, is easily transmitted down the covalently bonded backbone of a polymer but is not easily transmitted from one chain to another across van der Waals bonds. Polymers, therefore, show very different values of THERMAL CONDUCTIVITY in the direction of the chain as compared to other directions. Because of this consistent difference in transmitting ability of the inter-chain versus the intra-chain, the inter-chain energy transmitting ability is always the limiting aspect to the overall value of thermal conductivity. Furthermore, the thermal conductivity between chains is very sensitive to the distance of separation. Thus, conditions that serve to strengthen the inter-chain bonding and/or reduce the distance between chains serve to improve the thermal conductivity of the material. Polar side groups and crystallinity both increase thermal conductivity for this reason. Decreasing the temperature of a polymer also serves to decrease the inter-chain distance and increase the thermal conductivity. But, for most polymers, the change in thermal conductivity over their useful temperature range is less than ten percent, because the reduction in chain spacing produced by lowering the temperature will be about balanced by the decrease in vibrational energy available at the reduced temperature.

Additives can often have a large effect on the thermal conductivity of a polymer. Usually, the contribution of the additive is in proportion to its weight percentage. Low molecular weight organic plasticizers have a higher thermal conductivity than polymers due to their inherent mobility. Ceramics have a thermal conductivity about an order of magnitude greater than polymers; metals have a thermal conductivity about three orders of magnitude greater than polymers; and carbon-black characteristically displays a thermal conductivity value between that of a ceramic and a metal. Thus, virtually any additive serves to increase the thermal conductivity of a polymer, varying from a small to a large amount. This becomes a consideration in designs that use polymeric materials for thermal insulation. Often the best reinforcing agent for the polymer is also the worst with respect to changing the

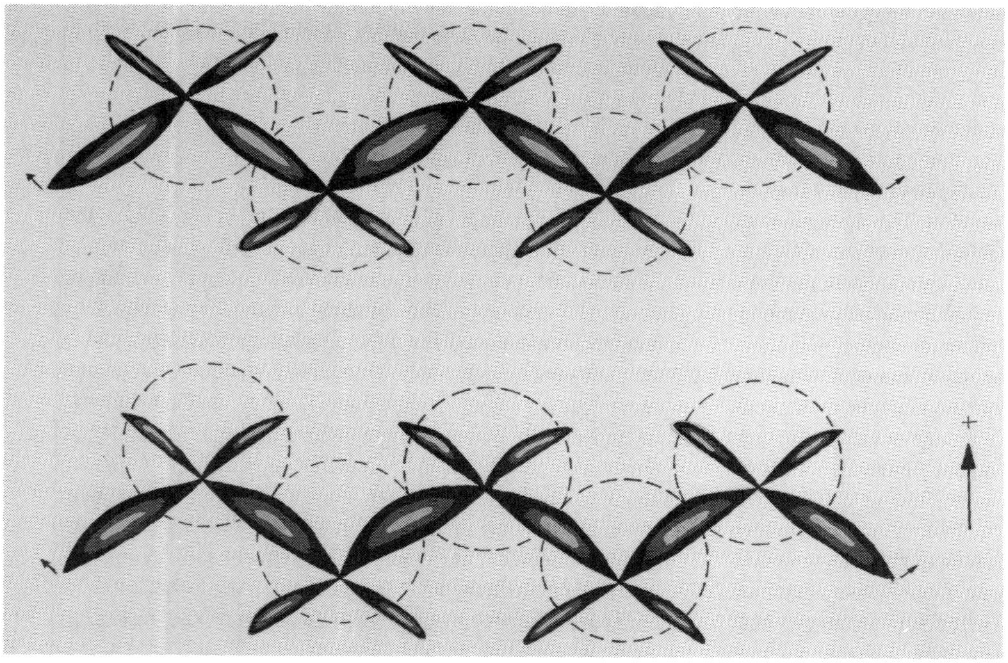

Figure 3.1. ELECTRON ORBITAL CHANGE DUE TO ELECTRIC FIELD
The upper schematic is that of a section of a polyethylene chain under normal conditions, with the small arrows indicating the continuation of the chain. The lower schematic is the same section of polyethylene chain with a positive electric field applied in the direction of the large arrow. Note that the *s* orbital is more obviously affected by the field since the *p* orbitals are considered to be the bonding orbitals in this schematic. It is this slight shift in electron orbital position relative to the normal centering about the nucleus that accounts for the electron contribution to the dielectric constant.

thermal conductivity. Thus, careful tailoring must be done to optimize the results.

In the special case of a polymeric foam, the additive is a gas which has the mobility to transmit energy by convection as well as conduction. As a result, thermal conductivity of a foam increases directly and linearly with cell size. Since the gas has very limited ability to absorb energy, its contribution to the transmission of thermal energy is low compared to that of the polymer. Consequently, the less polymer in a foam, the lower the thermal conductivity, up to the point of very thin wall thickness. Further improvement of the insulating ability of the foam can be effected by removal of the gas altogether, provided this does not cause collapse of the foam structure. This has been done for thin-walled, small sized, closed cell foams with dramatic results.

DIELECTRIC CONSTANT

The DIELECTRIC CONSTANT for polymers is lower than other types of materials and varies with structure. Polymers that do not have a dipolar structure and are free of other sources of dipoles display a consistently low value for the dielectric constant, which does not change with temperature or frequency by more than a small percentage. Even the glass transition temperature will leave this value almost unaffected if the polymer is purely covalent. However, any presence of dipoles can increase the dielectric constant by as much as an order of magnitude and make it a sensitive function of both temperature and frequency. Dipolar structure can be an inherent part of the polymer caused by the presence of an ionic component to the bonding, an ionic segment in the backbone, and/or ionic pendant groups. Dipoles can also be introduced via ionizable additives, impurities, and/or moisture.

When an electric field is applied to a polymer, the electrons are always free to respond to the field, but they may be the only part of the material that is free to respond to the field in a polymer with only covalent bonds or a crystalline structure. The electrons move in the direction of the positive side of the field and the positively charged atomic nuclei try to move toward the negative side of the field. The nuclei can not move effectively due to the restraint of the chain of which they are a part. The electrons do move a small distance but are prevented from leaving the vicinity of the nucleus to which they are bound by the restraining force of attraction to the nucleus. The net result is that the electronic orbitals are distorted in the direction of the positive electric field. An illustration of this can be seen in Figure 3.1. Since the electrons are small particles of low mass, they can move quickly in response to an electric field and are thus unaffected by frequency and temperature.

In amorphous polymers that contain a dipolar structure, there may be a contribution to the dielectric constant from the dipolar structure as well as from the electrons. For this to occur, the dipolar structure must have some mobility, so it can follow the changes in the applied electric field. At very low temperatures, dipolar structures do not have sufficient mobility due to their relatively large mass as compared to an electron. If the dipolar structure is in the form of a pendant group which is attached to the backbone via a flexible bond, then sufficient energy will eventually be present in the amorphous polymer on heating to reach a point at which the pendant group will be able to respond to the electric field. The temperature at which this occurs is usually well below the glass transition temperature. With increasing stiffness of the bond between the pendant and the backbone, the temperature of initial contribution of the pendant to the dielectric constant will increase. Thus, the range of frequency to which the pendant can respond is limited by its mobility, which is a function of both the energy needed to move its mass and the stiffness of the bond with the backbone. As the temperature is further increased, the mobility of the pendant continues to increase, and it would be expected that the dielectric constant would increase with it. However, increasing temperature also increases the thermal vibration of the atoms, allowing more random motion. These two factors tend to nullify each other, so that the dielectric constant is nearly uniform after the onset of pendant motion, until another form of dipolar mobility becomes available.

Another possible contribution to the dielectric constant comes from dipolar structures that are part of the backbone in amorphous polymers. These can only move in response to the electric field when the backbone has sufficient mobility to allow the segment in which they are located to move. This mobility is gained at the glass transition temperature. Above this temperature the segments of the amorphous polymer continue to gain mobility with increasing temperature, but, like the pendant motion, the random motion of the segment tends to nullify whatever contribution this added mobilty of the segment might bring about to increase the dielectric constant. Thus, the initial contribution that this structure makes to the dielectric constant at the glass transition temperature will remain about constant with increasing temperature.

Crystalline polymers behave similarly to amorphous polymers of the same composition. The contribution of the notion of electrons to the dielectric constant for the crystalline part of a polymer will be the same as in the amorphous form since electron motion is indepen-

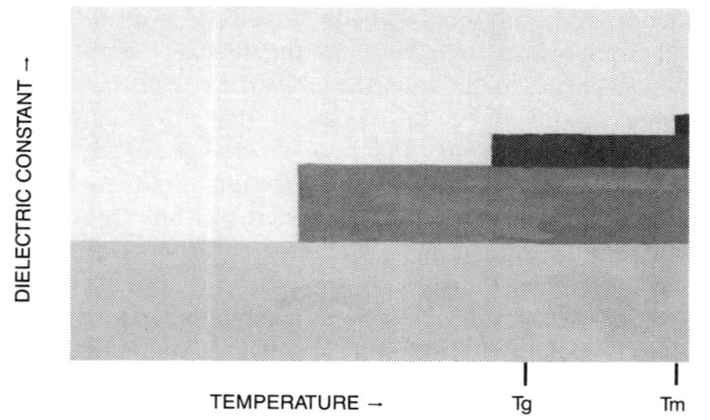

Figure 3.2. DIELECTRIC CONSTANT AS A FUNCTION OF TEMPERATURE

This diagram shows the sources of dielectric response by a polymer as a function of temperature, assuming constant frequency. Tg marks the glass transition temperature and Tm marks the melting temperature. Light gray represents the contribution of electrons to the dielectric constant and encompasses the entire temperature range of a polymer. Middle gray represents the possible contribution made by dipolar pendants. Dark gray represents the possible contribution made by dipoles that exist along the backbone of a polymer. The black region represents the possible contribution that dipolar structures could make in a crystalline polymer if the dipoles were able to rotate within the lamellae. A real polymer may have one or more of these contributions to its dielectric constant and each contribution is additive.

dent of structure. The contribution due to dipolar structure in the crystalline part of a polymer is usually negligible, since the dipoles are not free to move with the electric field. Thus, the dielectric constant of a crystalline polymer is about equal to the dielectric constant of the amorphous part of the polymer multiplied by the weight fraction of the amorphous part. However, the crystalline part can contribute to the dielectric constant due to its dipolar structure, once it has sufficient mobility to allow entire lines of the lammelae to rotate relative to the electric field. This occurs as the crystal approaches the melting point, providing the crystal is free of bulky pendants or branching which would prevent rotation. A chart of temperature related effects on the dielectric constant can be seen in Figure 3.2.

Since the dielectric constant depends on the amount of material available for transmission of energy, any process that reduces the density of the material will decrease the dielectric constant. Foams are particularly useful in this respect. They often have sufficient strength for design requirements while reducing the dielectric constant below that of the polymeric base of the foam. Conversely, the measurement of the dielectric constant can be used as an indirect means of determining the density of the foam.

DIELECTRIC BREAKDOWN

As long as the dipoles within the polymer are able to follow the changes in the electric field to which they are subjected the polymer shows a low LOSS FACTOR. However, whenever a type of dipolar motion becomes inoperable due to a change in conditions, the dielectric constant is reduced and the loss factor peaks. The peaking of the loss factor signals dielectric breakdown and corresponds to the beginning of physical breakdown in one of the types of dipolar response.

A change in temperature can cause loss factor peaking with its associated dielectric breakdown. As the temperature increases, various dipoles become mobile and contribute to the dielectric constant. If the reverse process (decrease in dielectric constant with decreasing temperature) is considered, then a loss factor peak and accompanying dielectric breakdown would be evident as each dipolar structure becomes too rigid to respond to the electric field.

The dielectric constant is also dependent on frequency. Each decrease in dielectric constant with increasing frequency can be associated with a loss factor peak and a dielectric breakdown. Figure 3.3 shows the response of both loss factor and dielectric breakdown in a frequency region where a dipolar structure is no longer able to respond to the electric field. Polymers that only have bound electrons with which to respond to the electric field are effectively immune to frequency dependence since their only dielectric contribution, the electrons, can follow an electric field up to a frequency of 10^{12} Hertz. However, other polar structures are associated with particles of larger mass than the electron and thus require more time to respond to the electric field. There is, then, a maximum frequency at which each polar structure can respond to the electric field, just as there is a minimum temperature for which each polar structure has sufficient mobility to respond to the electric field. The value of that frequency is also dependent on the mobility of the dipolar structure. Thus pendant dipolar structures with flexible bonds to the backbone not only have a wider temperature range within which to contribute to the dielectric constant, they also have a wider frequency range. A diagram showing temperature and frequency ranges for the contribution of each dipolar structure can be seen in Figure 3.4. Another representation of the dielectric breakdown points can be seen in Figure 3.5 which shows frequency versus dielectric constant at a single temperature. This latter figure is more easily interpreted if viewed in conjunction with Figure 3.2, which shows temperature versus dielectric constant at a single frequency.

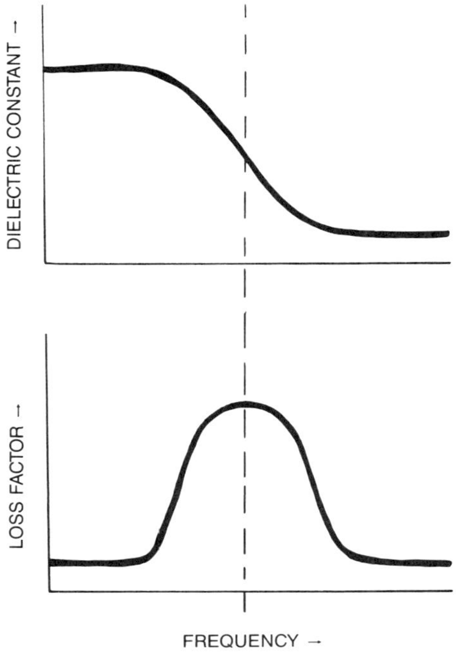

Figure 3.3. THE RESPONSE OF DIELECTRIC CONSTANT AND LOSS FACTOR IN THE VICINITY OF DIELECTRIC BREAKDOWN

These graphs show the typical reaction of a polymer to dielectric breakdown as it passes through the breakdown frequency, which is depicted by a dotted line. As the dielectric constant begins to decrease, there is a corresponding increase in the loss factor until the breakdown frequency is reached. At this point, both dielectric constant and loss factor decrease with increasing frequency until a new stable value is reached.

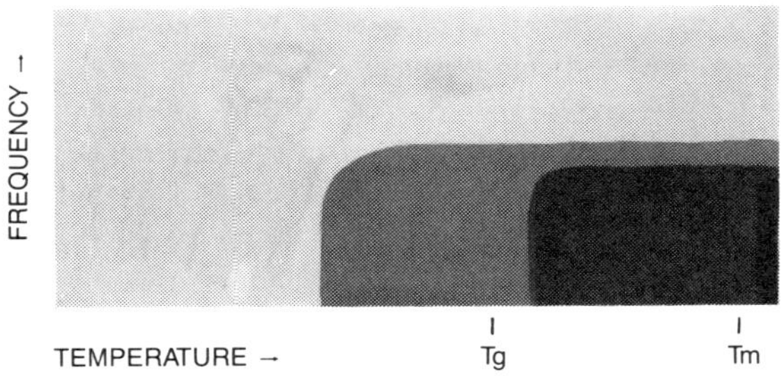

Figure 3.4. DIPOLAR MOBILITY AS A FUNCTION OF TEMPERATURE AND FREQUENCY

This diagram shows the types of dipolar structures that are mobile as a function of temperature and frequency for a typical polymer. As in Figure 3.2, light gray represents the bound electron dipoles, middle gray represents the mobile dipolar pendants, dark gray represents the mobile dipoles bound to the backbone, and black represents mobile crystalline lamellae which contain dipolar structures. Again, Tg marks the glass transition temperature and Tm marks the melting point. In this diagram, each successively darker region overlaps and includes lighter ones. Thus, the bound electrons are shown as contributing to the dielectric constant at all temperatures and frequencies.

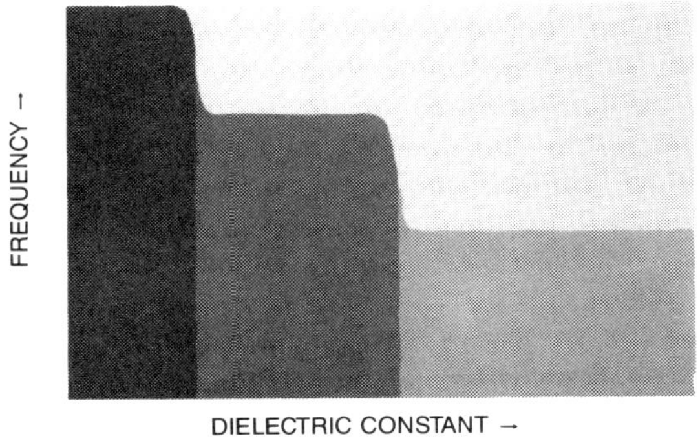

Figure 3.5. DIELECTRIC CONSTANT VERSUS FREQUENCY AT CONSTANT TEMPERATURE

In this chart the dark gray region shows the range of frequency over which the electrons, mobile pendants and mobile backbone bound dipoles are contributing to the dielectric constant. As the frequency increases, it reaches the limit of the dipolar backbone segment mobility and this contribution stops, resulting in a decrease in the dielectric constant. The next region, that of middle gray, represents the range of frequency over which the electrons and mobile pendants contribute to the dielectric constant. With further frequency increases, the dipolar pendants are no longer able to follow the electronic field and the dipolar pendant contribution stops, further reducing the dielectric constant. Subsequent increases in frequency are responded to by only the electrons, which are represented by the light gray area. Electrons continue to contribute to the dielectric constant up to extremely high frequencies. The possible contribution due to dipoles embedded in a crystalline structure is not shown in this diagram, since this occurs only over a very limited temperature range and is not used in most applications.

Dielectrics are also limited by the amount of voltage that they can withstand without electrical and chemical breakdown. The maximum voltage a material can withstand is reflected by its DIELECTRIC STRENGTH. This is usually measured at some specific time after the voltage is applied, because dielectric strength is also time dependent. Very short-term (less than ten minutes) dielectric breakdown is usually associated with a localized flaw or impurity and is not considered representative of the material as a whole.

If the polymer survives the initial voltage applied, then the most common mode of long-term dielectric breakdown is due to overheating of the polymer, which leads to chemical breakdown. As chemical breakdown occurs, the polymer produces retained gaseous products which further exacerbate the dielectric breakdown due to the very much lower dielectric strength of the gaseous product in comparison to the polymer. When a dielectric polymer is utilized in air, the air becomes a potential source of long-term breakdown. The electric field converts the air into ozone and other activated gases, which in turn attack the polymer. This form of degradation is particularly prevalent when the electric field is strong (high voltage).

Not only air but also moisture and dust from the atmosphere contribute to the dielectric breakdown of a polymer. Moisture always decreases the dielectric strength of a polymer because water is a small mobile molecule capable of penetrating most polymers and carrying with it a permanent dipole. The magnitude of the effect of moisture pickup on the polymer is highly variable due to both the nature of the polymer and its structure. Similarly, the presence of dust always causes some dielectric breakdown, but the magnitude is also dependent on the polymer and other conditions. The combination of moisture and dust can produce TRACKING. The most extreme cases of tracking result not only in dielectric breakdown, but also in spontaneous combustion.

The presence of additives in a polymer may serve either as an advantage or a disadvantage with respect to dielectric strength. Organic additives will inevitably be smaller molecules than the polymer and as such will introduce a higher concentration of chain ends. The ends of the chains are much more mobile than the internal segments of the polymer and are also more vulnerable to chemical breakdown. The result is that organic additives will be detrimental to the dielectric strength of a polymer, especially if they also encourage water absorption. Inorganic additives usually improve the dielectric strength of the polymer if they are stable at elevated temperatures, have a stable covalent bond, do not attract moisture and have a neutral interface with the polymer. Most of the strengthening effect comes about due to the inorganic materials ability to restrict segmental motion in the polymer.

ARC RESISTANCE

When electrical discharge occurs, a polymer is subject to an electrical current which it cannot easily dissipate. If sufficient energy is involved, the result can be the local breakdown of the polymer and the creation of a conduction path produced by local carbonization of the polymer. ARC RESISTANCE is a measurable quantity of resilience to such breakdown. It is dependent on the duration of electrical discharge, voltage, current, frequency of the field, and temperature. The arc resistance of a polymer can often be improved by the presence of ceramic additives which serve to interrupt the direct conduction path of a polymer even when the polymer has been carbonized.

STATIC CHARGE

The most common means of developing a STATIC CHARGE in a polymer is to bring its surface into contact with another surface and then separate the two surfaces. It the two surfaces are rubbed together during their contact to the point of causing ABRASION this will augment the static charge build-up. During the contact between the two surfaces the electrons of the respective materials are transferred with a net accumulation on one surface and a net decrease on the other. Thus, one surface will acquire a negative charge due to an excess of electrons and the other surface an equal positive charge due to a deficit of electrons. Both surfaces are considered to be equally statically charged, since static charge is independent of charge sign and both surfaces will display the same static charge characteristics. Although static charge is usually considered a surface phenomenon, it is possible for a static charge to be induced into the bulk of the polymer. This is a relatively rare occurrence and still retains the same characteristics of the surface charge except in very specific electronics applications.

Because polymers are basically covalently bonded, they have little ability to dissipate or conduct electrons as needed to return the material to electrical neutrality after static charge has been established. Thus, polymers are able to retain a static charge over a long period of time, even as long as several years. It is this retention of the static charge which makes polymers unusual, as all materials develop static charges. Because of this capacity, polymers have been effectively used as electrostatic power coatings. The disadvantages of this retention ability range from the annoyance of dust attraction to the hazard of sparking. Static discharge can be particularly damaging to semiconductor devices and integrated circuits. Reduction of the ability of a polymer to create and retain a static charge can be accomplished by the use of anti-static additives.

ELECTRETS

ELECTRETS are an example of the application of static charge retention in polymers. They are made from materials that hold a static charge very well and have a glass transition temperature above the desired service temperature. An electret is made by heating a polymer to above its glass transition temperature and then subjecting it to an electric field that is maintained as the polymer is cooled. The electric field causes all of the ionizable particles to ionize, and the mobility of the rubbery region allows the ions to migrate towards their respective poles to a greater extent than would be possible in the glassy region. This migration will not be effective for bound ions like pendant groups, but will cause noticeable separation for impurities and additives that are able to dissociate. Once the polymer is cooled below the glass transition temperature, the ions are restricted in their mobility, and the electric field may be removed leaving the ionization of the dissociated impurities and additives retained. Over time the ions gradually recombine, so that an electret is in a slow but steady state of decline from its creation. This decline is usually uniform throughout the bulk of the polymer and can be well predicted for design uses. Since an electret is destroyed by heating it above its glass transition temperature, it can experience catastrophic failure. To minimize this problem, crystalline polymers have been successfully used as electrets.

PIEZOELECTRIC POLYMERS

Only wood and other cellulose-related materials fulfill the usual model of a piezoelectric material, since they are crystals. However, many other amorphous and semi-crystalline polymers exhibit the main property of PIEZOELECTRICITY: the generation of an electric current during or after deformation. Deformation causes the bonds in many polymers to be distorted such that a dipole is produced. Once the distorting force is removed the dipoles are free to de-polarize and a current is produced as a result of the depolarization. In most materials de-polarization is accomplished very quickly, but some polymers can retain the dipoles for months before dissipating them. As with electrets, this effect can be "frozen" into a polymer by deforming it above the glass transition temperature and releasing the force when the polymer has cooled to below the glass transition temperature, where recombination mobility is low. Also like electrets, dipolar recombination is slow but constant with time, and heating it to the glass transition temperature will dissipate the dipoles quickly.

The heating problem has been of concern for applications which use a wide range of temperatures. Currently the main use of this is in the electronics industry for audio equipment subject to electrical overheating. Piezoelectric polymers may also be inadvertently produced by the normal processing of a polymer into a final product. The majority of applications will be insensitive to this, but electrical applications may require the annealing of the polymer to remove the induced dipoles.

ELECTRICAL INSULATORS

Polymers are insulators unless very specific structures and/or additives are present. Polymers with the most nearly covalent bonding are the best insulators, as they are also the best dielectrics, since no carrier is really free to conduct electricity. Superb insulation has been obtained with polymers that are both covalently bonded and contamination free. Of course they are very difficult to prepare and maintain, so for normal use all polymeric materials do conduct some measurable amount of electricity.

The more ionizable a polymer the greater will be its response to an electric field. Dipoles induced in bonded atoms can contribute to the dielectric constant, but they have limited mobility and are therefore not very effective charge carriers. Thus, the polymer remains an effective insulator. However, the bonded dipoles within a polymer can attract mobile dipoles which will be free to conduct. An example of this is the attraction of water into a polymer. The more polar in nature a polymer, the more soluble will be the ionized water molecules from the atmosphere. Water can greatly reduce the resistivity of a polymer due to its additional charge and mobility.

Any other factor that contributes mobile ions to a polymer will also decrease the insulating ability. Increasing the temperature of a polymer increases its inherent mobility and also may serve to free some bound ions. This would be especially true above the glass transition temperature. Adding ionizable additives or impurities is an obvious source of ions not bound to a polymer. Reducing the pressure on a polymer will also decrease its resistivity for the same reasons that increasing the temperature lowers resistivity. Conversely, electrical resistivity measurements

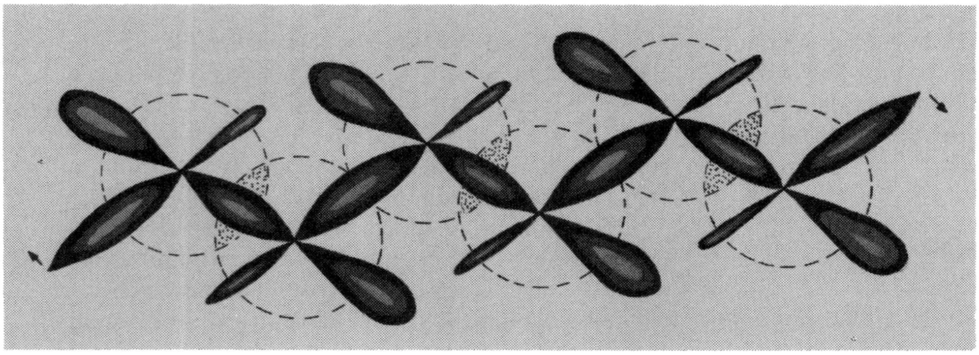

Figure 3.6. A CONJUGATED CHAIN OF POLYACETYLENE
Shown is a schematic of part of a polyacetylene chain. The arrows indicate the continuation of the chain. The structure is:

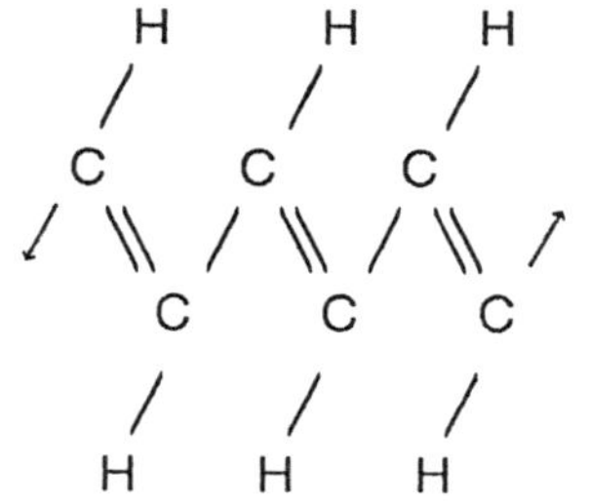

Double bonds alternate with single bonds along the backbone. These bonds also resonate so that the double bonds shift position regularly. This shift in bonding electrons could also be accompanied by the conduction of bonding electrons down the backbone rather than just a random shifting, if there were an electrical field to induce the motion of the electrons in a defined direction. Thus, this structure is capable of effective electron conduction down the backbone. This does not aid in the transfer of electrons between molecules.

are often used as a nondestructive means of evaluating a polymer for purity and structural flaws.

ELECTRICAL CONDUCTORS

Although polymers are most often used for their insulating ability, there are many areas in which a polymeric semi-conductor or conductor would be of great usefulness. The absorption of water vapor into an already polar polymer has produced a material approaching a semi-conductor, but this is not an easily maintained condition. More stable additives such as metal powder have been combined with polymers in an effort to make a conductive material. Some success has been achieved in this way but only with such large amounts of metal that the result is a composite. A more promising route for the production of a conductive material that still retains the characteristics of a polymer is in the direction of structural alteration.

One means of structural change is the irradiation of polymers. Irradiation disrupts the normal bonding structure and liberates charge carrying particles, which are free to move from the molecule. These charged particles will eventually recombine to form a neutral molecule but during their unbound period of time they are very effective in conducting electricity. The time until the recombination of the charge carrying particles and the ceasing of the electrical conductivity is difficult to predict and very sensitive to temperature. By means of irradiation a few semi-conducting polymeric materials have been produced, which fit into the usual design criteria for temperature range and useful lifetime.

Another means of producing a polymeric material which has a structure conducive to electrical conduction is to begin with a promising monomer and construct one by chemical reaction. There exists a whole family of polymers that are inherently semi-conductive and all have the same structural feature, a CONJUGATED DOUBLE BOND. A conjugated double bond forms a series of alternating double bonds down the backbone of a polymer, which serves as a passageway for charge conduction during resonance of the bonds. Figure 3.6 shows a schematic of the bonding.

Figure 3.7. LADDER STRUCTURE CONDUCIVE TO ELECTRICAL CONDUCTION

In this schematic, part of a polymeric backbone, which is composed of carbon rings, is shown. It has a structure of:

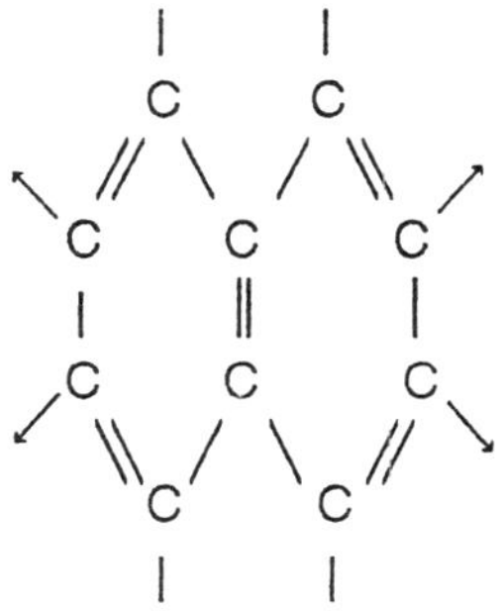

The arrows show the next bonding for the chain continuation. At first glance this structure appears as a normally bonded carbon chain with alternating double bonds that are free to resonate. However, only the four carbon atoms shown at the top and bottom peaks of the structure are in normal bonding configurations for organic carbon. These are:

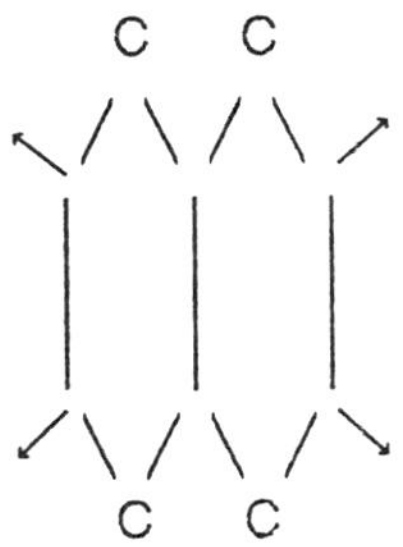

The other carbon atoms, which are shared at the joining of the rings, are in a hybrid state reminiscent of graphite. Thus, as the structure begins to approach that of graphite, the properties of the material approach those of graphite too.

With an established path through the polymer, it chain-end proximities in specific directions. Conductivity is reduced down the backbone if the backbone seems possible that a polymer with conjugated double bonds could be a conductor. Until recently, only semi-conductors had been produced from this structure. The introduction of metal and near-metal ions to conjugated polymers has successfully depolarized the electronic structure, permitting metallic levels of conductivity to obtain. The limiting factor remains the transfer of a charge from the end of one chain to the beginning of the next. The more alignment a polymer has, the easier this can be accomplished. As a result, aligned filaments and crystalline structures show an increased conductivity over the same polymer is an amorphous condition. This improvement is highly directional because both ordered structures favor chain-end proximities in specific directions. Conductivity is reduced down the backbone if the backbone twists the even array of alternating bonds out of alignment. Filaments and crystalline structures also tend to minimize twisting.

A refinement of the conjugated bond structure is the introduction of ring structures into the backbone of the polymer. The rings have resonating double bonds within them which act in the same way as a simple resonating double bond in the conduction of electricity. The use of the ring structure helps to prevent rotation of the backbone and the accompanying loss in its conducting ability. Figure 1.11 illustrates this type of structure.

A further refinement of the conjugated bond structure is the ladder-type backbone. In the ladder-type structure, rings are joined together such that two atoms of a ring are shared at each joining point. The resulting structure looks rather like a ladder. This structure is illustrated in Figure 3.7. As a consequence of the ladder structure, the polymer has a very rigid backbone that will not allow rotation. Thus, this structure facilitates conductivity and produces a high melting point and it also imparts more rigidity than would be expected from most polymers.

Additional gains in conductivity have been made by extending the ladder structure into a planar form similar to that of graphite. This is accompanied by further increase in the rigidity of the polymer rendering it useless for many mechanical applications. In addition, this structure produces exceptionally high dielectric constants which have been explained by the normal presence of unbalance charge carriers. These unbalanced charge carriers do not seem to recombine over time. They thus act as a constant source of conductivity within the polymeric chain. While they add greatly to the dielectric constant, they do not add significantly to the conductivity since such charge carriers still have the difficulty of inter-chain transfer.

PHOTOCONDUCTION

As with other energy sources, e.g., irradiation, exposure of a polymer to light causes some of the energy to be absorbed, allowing the liberation of charge carriers. When the light source is in the ultraviolet range the most common carrier liberated in a polymer is an electron that has been energetically elevated into the CONDUCTION BAND. Once the charge carries are free, they are able to conduct electricity. Usually the charge carriers will quickly recombine but if the light source continues to produce more charge carriers then a dynamic equilibrium will be reached. This phenomenon of electrical conductivity induced into a polymer via a light source is PHOTOCONDUCTIVITY. The effectiveness of a light source depends on the ease with which charge carriers are liberated from the polymer and the duration of time that will elapse before recombination occurs. In general, insulating polymers will have rapid recombination rates, whereas semi-conducting polymers will be quite sluggish in recombining.

LUMINESCENCE

LUMINESCENCE is almost the opposite of photoconduction. Photoconduction uses light as the energy source and produces charge carriers free to conduct electricity. Whereas, luminescence produces visible light when electrons that have been elevated to the conduction band by an energy source return to the valence band. Luminescence has no specific energy source responsible for the creation of free charge carriers. Thus, photoconduction only uses light as a source, and luminescence only produces light as a product.

The bundle of energy that the electron requires to escape to the conduction band or releases on returning to the ground state is a PHOTON. The photon emitted by the electron is characteristic of the atom to which the electron belongs, so carbon based polymers will all release a photon with a similar wavelength from their carbon atoms. Analogously, silicon based polymers will release photons with similar wavelengths and so on with the other elements. The light elements, which comprise most polymers, happen to emit photons with wavelengths within the visible light range. It would seem that such a light source would have very well defined wavelengths for each element

Figure 3.8. A SMOOTH POLYMERIC SURFACE REFLECTING LIGHT
In this diagram the gray represents the polymer which has a smooth surface. As the light rays, depicted by the heavy black lines, strike the surface of the polymer some of the light is reflected, depicted by the narrow black lines. It can be seen that the angle of the incoming light is equal to the angle of the outgoing reflected light but the direction is changed. The smooth surface produces a consistent angle for the reflected light; hence, it appears shiny.

and thus produce a pure monochromatic light. However, this is only true for very uniformly bonded materials which are free from impurities. Any weak point in the bonding of a polymer will be more likely to respond to the energy source and propel electrons into the conduction band than the average bond of a polymer. Thus, double bonds, chain ends and impurities will be the most likely sites for an electron to gain promotion into the conduction band. As a result, luminescence is much more likely to reflect the presence of double bonds and small molecular weight impurities than the nature of the large chain.

The specific terms used within the general category of luminescence are based either on the response time of the polymer to producing light after it has been exposed to an energy source or the nature of the energy source. If the recombination and emission of the photon of light occur immediately after the energy source is removed then the polymer is called FLUORESCENT. If the recombination is a slow process taking place with a decaying rate then it is said to be PHOSPHORESCENT. When luminescence is based on a specific type of energy source then the name clearly reflects this. Four of these terms are THERMOLUMINESCENCE, ELECTROLUMINESCENCE, TRIBOLUMINESCENCE, and CHEMILUMINESCENCE.

OPTICAL PROPERTIES

Light is REFLECTED to some extent by all materials and a smooth surface encourages the light to be reflected in a simple manner resulting in a glossy appearance. Light that is reflected is not transmitted through the material and cannot emerge at the other surface of the material. The total of all light that does not emerge from the other surface of the material is said to be ABSORBED. This includes light reflected at the surface, light absorbed by the material, and light that is reflected within the material and ultimately is either absorbed or transmitted through the material at another surface.

Polymers as a class of material can display very similar optical properties. They can be clear, glossy and colorless in appearance, under at least some conditions. The clarity comes about due to the completely bound nature of the electrons. With the electrons all associated with specific atoms and tightly bound in defined orbitals, the polymer has a great deal of "open space" through which a photon of light may pass without "striking" a particle. The clarity of a polymer can be altered in two ways: by surface roughness and internal voids. The creation of surface roughness will also change the surface finish to a matte appearance.

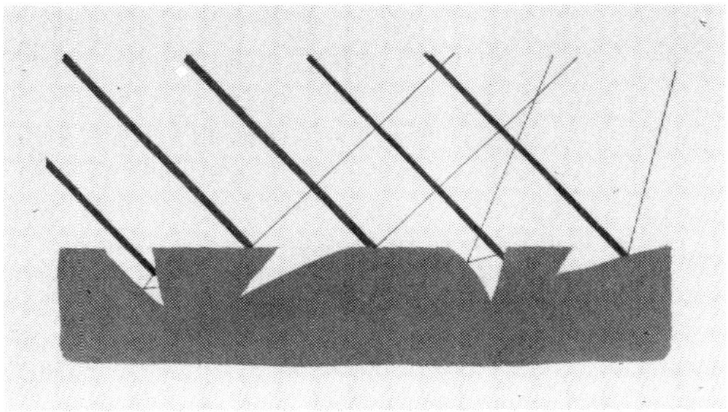

Figure 3.9. A CRAZED POLYMERIC SURFACE REFLECTING LIGHT
This diagram is the same as Figure 3.8 except the surface in this one has a series of crazes or cracks in it. When the light hits a crack it either reflects within that crack and fails to return a reflected ray or the reflected ray is at a different angle from those that are emanating from the smooth portion of the surface. Note, however, that the angular relationship between the incoming light and the reflected light remains the same. Since the reflected light no longer has a well-defined angle relative to the perceived surface of the polymer, the result is a cloudy appearance.

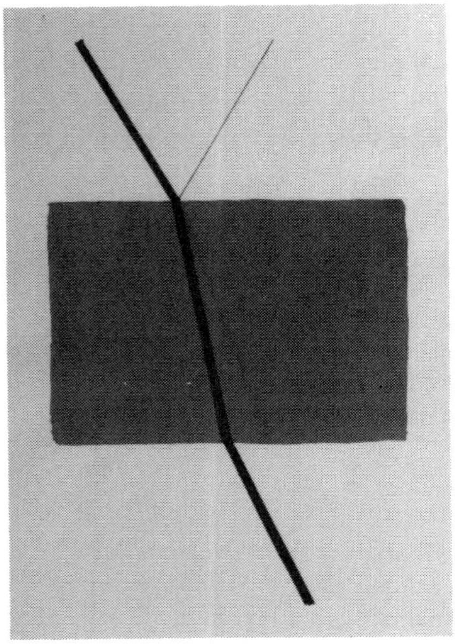

Figure 3.10. REFRACTION OF LIGHT IN A POLYMER
The polymer in this diagram is depicted by the gray area. The light beam is shown as a wide black line and the reflected light is shown in more narrow lines, the thickness of which roughly corresponds to their intensity. As the light beam strikes the surface of the polymer some of the light is reflected and some is refracted into the polymer. The light emerging from the polymer is refracted again. There is also a light reflection from the second surface of the polymer which is not shown.

Coloration occurs in a material if the elements that comprise that material are able to absorb photons of light and re-emit them in the visible range while absorbing some of the energy associated with the wavelength of the color it will display. Polymers are comprised of lightweight elements which do not absorb or re-emit energy in the visible range even when they may be struck with visible light. Color is almost always introduced by inorganic additives. Most additives for strengthening or other non-decorative purposes are either white or black. These may be difficult to mask if another color or translucency is desired for the finished product.

REFLECTION

Polymers are capable of having very smooth surfaces, and little light will be lost to reflection on this type of surface. Polymers are also subject to surface flaws which can develop during processing or in service. They are easily scratched, abraded, corroded, and degraded. If the surface has flaws from CRAZING and FROSTING, then the incoming light is both reflected and scattered at the surface resulting in a translucent matte appearance. Figure 3.8 shows reflection for a smooth surface and Figure 3.9 shows it for a rough surface.

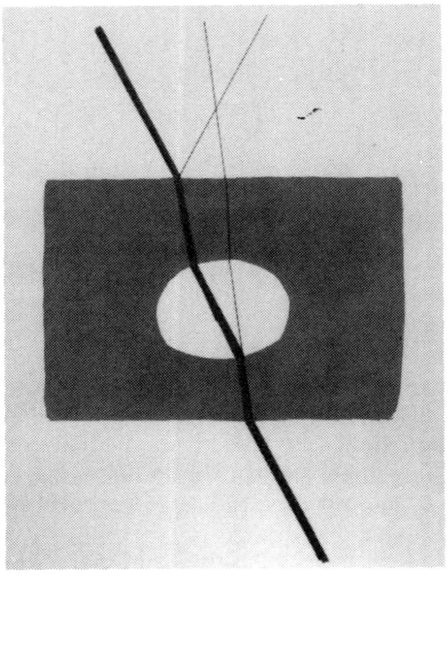

Figure 3.11. REFRACTION OF LIGHT IN A POLYMER WITH A VOID
As in Figure 3.10, this figure shows light passing through a polymer. However, in this figure the light also passes through a void in the polymer which both refracts and reflects the light. Only the primary reflection of the light in the void is shown for clarity. If this single reflection is followed, it can be seen that the reflected light emerges from the top surface of the polymer at a very different angle from that of the reflected light of the top surface. This same behavior is displayed by the cells in a polymeric foam.

REFRACTION

As light is transmitted through the surface of the polymer, all except that which is reflected becomes REFRACTED by the polymer. The amount of refraction is dependent on the density of the material through which the light passes and can be calculated using the index of REFRACTION and Snell's law. Upon emergence from the polymer, the light refracts again in accordance with the index of refraction of the new medium through which it is passing. An illustration of this can be seen in Figure 3.10. If a polymer has a void, then the surfaces of the void will reflect light in a similar manner to the surface of the polymer. The interior of the void will also have its own refraction index which will probably be that of the atmosphere surrounding the polymer. Because of the combination of a differential refraction of the light by the polymer versus the void and the differences in surface alignments of the polymer versus the void, some of the light passing through the void will be reflected and refracted away from the path of the main beam of light traveling through the polymer. Figure 3.11 illustrates this. Reflections and refractions similar to that of the void occur if the light beam hits an additive or an impurity. These affect the light beam in a manner consistent with their own optical properties. Most likely they will absorb the light which strikes them.

ABSORPTION

All of the ways in which the light beam is reflected, absorbed or refracted away from the main beam of light emerging from the polymer are considered losses to the transmitted light and are collectively referred to by the general term *absorption*. As the absorption is increased, the polymer goes from transparent to translucent and then finally to opaque. The means of increasing absorption in a polymer are dependent on the possibilities for light scattering and light absorption without re-emission. Polymers are not light absorbers, so only additives can be a source of absorption without re-emission. However, disruptions in the polymeric structure can cause light scattering to such an extent that a polymer may become opaque even without additives to assist in absorption. An example of this is the addition of voids to the polymer to such an extent that the polymer becomes a foam. As would be expected, the greater the number of cells or voids per centimeter of thickness, the greater the opacity of the polymer. This is usually independent of the size of the cell or void and the thickness of the polymer between the void.

Another structure which usually increases absorption within a polymer is crystallinity. Polymers with differences in density between the crystalline and the amorphous forms have different refractions for each region. As the light goes from one density region to another its refraction changes. Accompanying the refraction change is surface reflection. Thus, each tiny region of crystalline and amorphous polymer acts like its individual constituents, with a great deal of opportunity for reflection from all of the surfaces available. Although the reflection is slight at these surfaces, the number of them makes the overall contribution significant. In addition, there is mismatch of structure at the boundaries of the spherulites, which contributes to the reflection. As a result, polymers with a difference in density of the crystalline versus the amorphous regions will be less transparent than the same polymers with a completely amorphous structure. The greater the difference in density between the crystalline and amorphous regions, the greater the opacity. A further consideration is the size of the spherulite. Large spherulites reduce the amount of boundary mismatch within the polymer and improve transparency over smaller spherulites. Thus, annealing of the polymer to encourage spherulite growth is a means of improving transparency, except when the spherulites have failed to fully form in the polymer. If this is true, annealing will only encourage additional crystal formation and reduce transparency.

ACOUSTICAL PROPERTIES

Sound can be reflected, transmitted or absorbed by a polymer in a similar fashion to light. In general, sound reflection is small and can be further reduced by the use of open-celled foam which can almost eliminate REVERBERATION. Transmission of sound through a material is inversely proportional to the density. Thus, polymers are very effective sound transmitters. Foams further reduce the density and therefore are even better transmitters than their basis polymer. Conversely, crystalline polymers usually have a greater density then the amorphous phase. Consequently, crystalline polymers absorb sound more readily than those in the amorphous phase.

Sound absorption is not a linear function of density. In fact, polymers absorb sound at very specific frequencies, which are characteristic of the polymeric composition and structure. Since these are rather pointed and narrow peaked functions, polymers have been used as filters for specific frequencies. Usually, this is done with a series of polymers with different absorption characteristics to remove such unwanted sound as background noise.

Tensile Properties 4

TENSILE TEST

One of the most extensively used mechanical tests for polymers is the TENSILE or STRESS-STRAIN test. The test is usually done at a specified rate and sample configuration that are similar for all polymers. The results are widely different for different polymers and can vary greatly for the same polymer. Such variation in test results is primarily due to the diversity of structures found within the broad range of polymers. Similarly, the YOUNG'S MODULUS, FLEXURAL MODULUS and FLEXURAL STRENGTH are highly variable with varying test conditions, often with unexpected results. Schematics of tensile curves for different types of polymers can be seen in Figure 4.1. When a family of such curves is examined, it is apparent that polymers are ANISOTROPIC and highly sensitive to both strain rate and temperature. Many polymers show no linear region that could be interpreted as an ELASTIC REGION, and when such does exist the associated Young's modulus will be small. Furthermore, the COMPRESSION results are generally significantly different from the tension results.

Tensile values expected for a polymer do have ranges which can be useful to remember. The tensile strength and YIELD POINT both vary between about 7 and 70 MPa (1,000 and 10,000 psi). Strain-to-failure varies from 1% to 1000% when different types of polymers are considered, but the variation is somewhat less if the same polymer is examined. Of course these values can vary further with the use of additives.

NECKING

Almost all polymers that are not cross-linked will NECK under some conditions during testing. Necking is manifested by a physical change in the specimen such as that depicted in Figure 4.2. Along with the physical change in the geometry of the specimen there is a corresponding change in the slope of the tensile test graph, which is shown in Figure 4.3. While the polymer is necking, the same stress will cause considerable elongation, so that the necking region of the tensile test graph is virtually a straight line parallel to the strain axis. Once the specimen has necked as much as it can, the specimen cross-section will usually be uniformly thin. Further strain increases will cause the stress to increase too, and the specimen will break shortly after necking ends.

Just prior to the onset of necking, the tensile test graph shows a small stress maximum, or hump, which becomes more pronounced as the testing rate is increased. It is also possible for the hump to be eliminated if the test is done slowly enough. Necking is always accompanied by localized heating and it is felt that this heating effect causes the hump. The reasoning is that faster testing would allow less time for the heat to dissipate and augment the heating effect. The heat would cause a softening of the polymer causing the strain to drop a bit rather than be maintained at the same level throughout necking. Finally, necking has useful applications in the formation of FIBERS and in COLD DRAWING.

THE EFFECTS OF TEMPERATURE ON TENSILE PROPERTIES

As temperature increases in a material there is a gradual expansion of the material, resulting in more free volume and a weakening of the bonding forces that hold the material together. In polymers this is mainly manifested in a reduction in the van der Waals forces between molecules. As such, all materials have less internal strength with increasing temperature, which is reflected in reduced maximum STRENGTH. Accompanying the reduction in strength is an increase in the strain which the polymer can sustain. This is illustrated in Figure 4.4 which shows a family of tensile graphs for a polymer within its DUCTILE region.

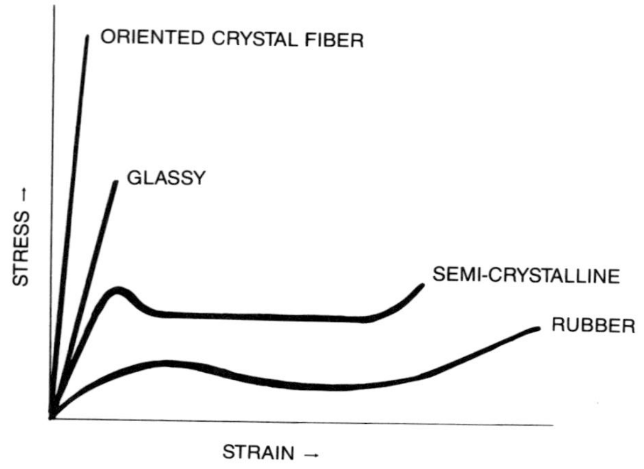

Figure 4.1. SCHEMATIC OF TENSILE TEST RESULTS

The four tensile test results given here are typical of polymers in different states that have been tested to failure with a moderate strain rate. The first curve to the left shows the extreme case of a crystalline state with the lamellae oriented in one direction such as would occur within a fiber. This curve is also characteristic of a highly cross-linked polymer. The next curve shows typical behavior for a polymer in the glassy state. The third curve is that of a semi-crystalline polymer. Further details on this type of curve are given in Figure 4.3. The last curve is that common to the rubber state which shows no real linear region at the onset of testing.

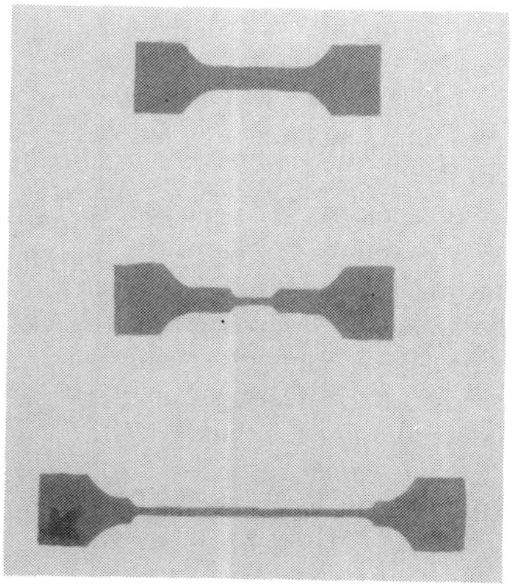

Figure 4.2. NECKING OF A TENSILE SPECIMEN

The top drawing shows a polymer tensile specimen which has yet to be tested. The middle drawing shows this same specimen after the onset of necking. The bottom drawing shows the specimen after necking is complete.

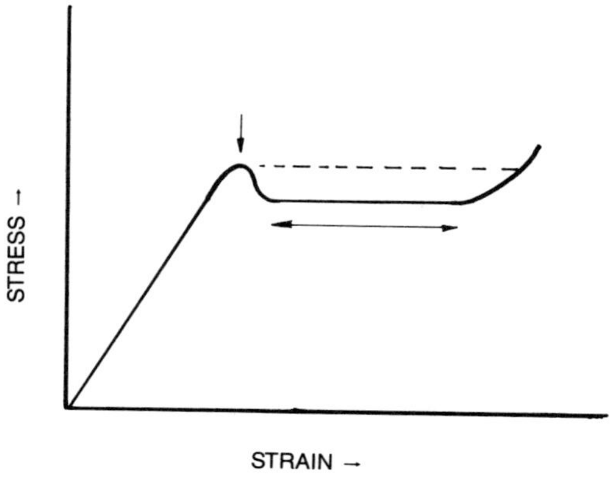

Figure 4.3. TENSILE TEST GRAPH FOR A TYPICAL CRYSTALLINE POLYMER

This graph shows the usual tensile test results for a partially crystalline polymer. The first section is basically elastic and the stress-strain relationship is linear. A hump, indicated by the single-headed arrow, results for moderate strain rates and signals the onset of necking in the specimen. Necking continues in the region marked with the double-headed arrow. At the end of this linear region, necking is completed and the specimen breaks shortly thereafter. If a slow strain rate is used in testing, it is possible to avoid the hump in the curve, and the tensile test results will then follow the dotted line. The necking will start and end at the same strain in the slower testing as in the moderate testing.

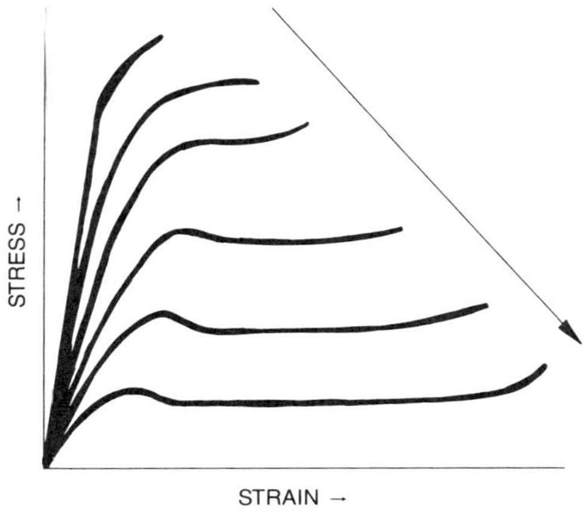

Figure 4.4. VARIATION OF STRESS/STRAIN RESULTS WITH TEMPERATURE
This graph represents a family of tensile tests taken for a single polymer over its ductile region and at a moderate testing rate. The tests were each conducted to failure at a single temperature. The difference between successive testing temperatures is about 20° Kelvin. The arrow indicates the direction of increasing testing temperature. In comparing this graph to Figure 4.1 it is evident that the basic form of this material corresponds to the third graph of Figure 4.1, that of a semicrystalline polymer. The graphs of this figure vary from approaching the glassy behavior on the low temperature end, to that of the rubber state on the high temperature end.

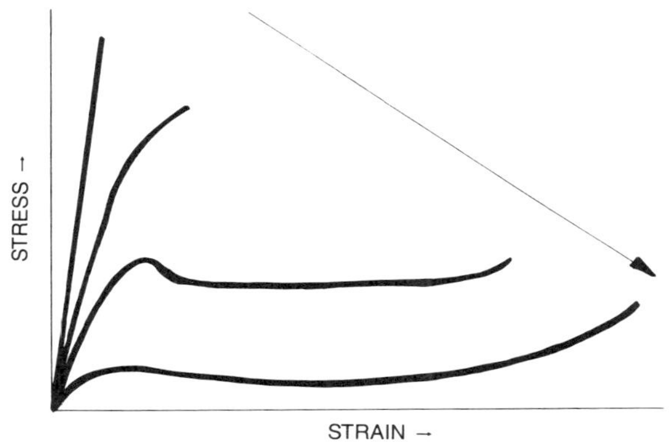

Figure 4.5. TENSILE BEHAVIOR AS A FUNCTION OF TEMPERATURE
As in Figure 4.4, this graph shows a family of tensile tests taken for a single polymer, but this graph covers the full range of tensile behavior. Each curve represents a test conducted at a single temperature and the arrow indicates the direction of increasing temperature. The curve farthest to the left was obtained in the glassy region below the glass transition temperature. The remaining three curves were obtained above the glass transition temperature. The second curve from the left shows limited ductility, the third curve shows increased ductility with a necking region and the fourth curve shows true rubber behavior. This graph can be compared to Figure 4.1 in which the curves were all obtained at the same temperature but the structure and/or polymer were varied.

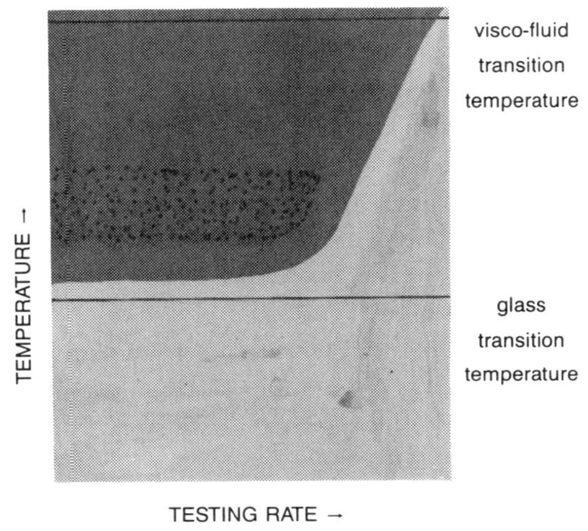

Figure 4.6. SCHEMATIC OF REGIONS OF DUCTILE AND BRITTLE BEHAVIOR
This schematic shows the region of expected brittle behavior (light gray area), versus the rubber or ductile behavior (dark gray area), as both a function of temperature and testing rate. The glass transition and visco-fluid transition temperatures are noted to give the relative relationships of these to the expected mechanical behavior of a polymer. The dotted region indicates the conditions under which necking would be expected to appear.

Besides the effects of thermal expansion, an increase in temperature brings about changes in the state of the amorphous polymer. In conjunction with these changes of state are changes in the tensile behavior of the polymer. At the lowest temperatures polymers are BRITTLE. As the temperature increases they become more TOUGH, until they reach the BRITTLE-DUCTILE TRANSITION. It is above this transition temperature that polymers become sufficiently ductile that they can exhibit necking. Still further increases in temperature produce a rubber-like behavior that is displayed until mechanical strength breaks down. A schematic of these tensile graphs can be seen in Figure 4.5.

Although there is a relationship between mechanical properties and structure, the transition temperatures for structural changes do not correspond to the transition from one mechanical response type to another. As an example, rubber-like tensile graphs are obtainable from polymers in the rubbery region above the glass transition temperature, but the onset of this type of mechanical behavior is well above the glass transition and continues into the visco-fluid region, provided slow to moderate strain rates are used. If high strain rates are used, the polymer can be made to show brittle behavior at almost any temperature. See Figure 4.6 for a schematic of these regions of behavior.

Crystalline polymers follow the mechanical behavior of amorphous polymers at the low temperature end. Their brittle-ductile transition temperature, as their glass transition temperature, is established by the amorphous content of the material. Thus, the more crystalline the material is, the less sensitive it is to the changes which are brought about by the amorphous content. Crystalline polymers always show higher strength and less DUCTILITY than the same polymers in the amorphous state at the same temperature. At the high end of the temperature scale, even a relatively small crystalline content will prevent a polymer from displaying rubber-like tensile properties. Crystalline polymers usually continue to follow the semi-crystalline behavior shown in Figure 4.1 up to the CRYSTALLINE MELTING TRANSITION TEMPERATURE. At this point the mechanical properties deteriorate precipitously.

Cross-linked polymers have the same range of mechanical properties displayed by an amorphous polymer, so long as the cross-linking is light. As the cross-linking increases, the mechanical properties are modified, such that at the extreme end of high cross-linking, the polymer is virtually insensitive to temperature changes. Examples of the variation of Young's modulus with temperature for the various types of polymers can be seen in Figure 4.7.

THE EFFECT OF STRAIN RATE ON TENSILE PROPERTIES

The most common testing rates span several orders of magnitude, from about .001 to 1 cm/sec. With increasing strain rate, the strength and Young's modulus increase with a corresponding decrease in elongation. Similar trends are found for both tension and compression testing, although a simple relationship between the two types of tests is not obvious. The change in tensile behavior with increasing strain rate is also similar to the tensile behavior of a polymer with decreasing temperature. By varying either temperature or strain rate, it is possible to exhibit tensile behavior from brittle through ductile for most polymers. When both strain rate and temperature are varied, the two effects may compensate for each other,

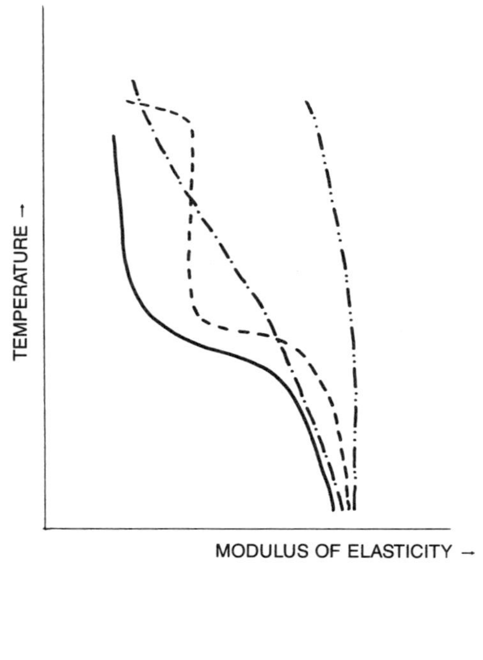

Figure 4.7. THE RELATIONSHIP BETWEEN TEMPERATURE AND MODULUS OF ELASTICITY FOR VARIOUS STRUCTURES

This schematic shows the variation of the modulus of elasticity with increasing temperature for four different structures. The solid line represents a typical amorphous polymer. The "knee" or plateau in the curve approximately corresponds to the glass transition temperature. The dashed line represents a typical crystalline polymer. This structure has a higher yield strength than the amorphous structure for the same polymer, but the variation of the modulus of elasticity with temperature follows the same form as that of the amorphous polymer since the amorphous component of the crystalline polymer undergoes a glass transition which the crystalline component does not. The dashed line with a single dot between the dashes is representative of a lightly cross-linked polymer. The cross-linkage makes the polymer relatively insensitive to the glass transition temperature. Further cross-linking results in a curve like the dashed line with two dots between the dashes. Highly cross-linked polymers are relatively insensitive to both glass transition and increased temperature.

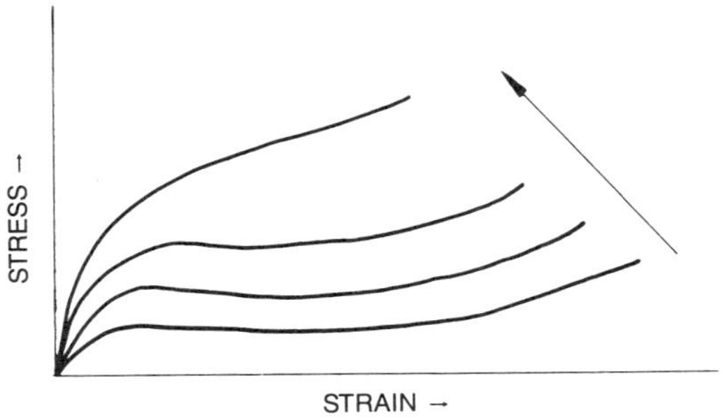

Figure 4.8. THE EFFECT OF STRAIN RATE ON TENSILE BEHAVIOR

This is a graph showing the general effect of strain rate variation on the tensile behavior of a polymer with all other variables being held constant. Each curve shows the tensile test done at a constant strain rate with the arrow indicating the direction of increasing strain rate.

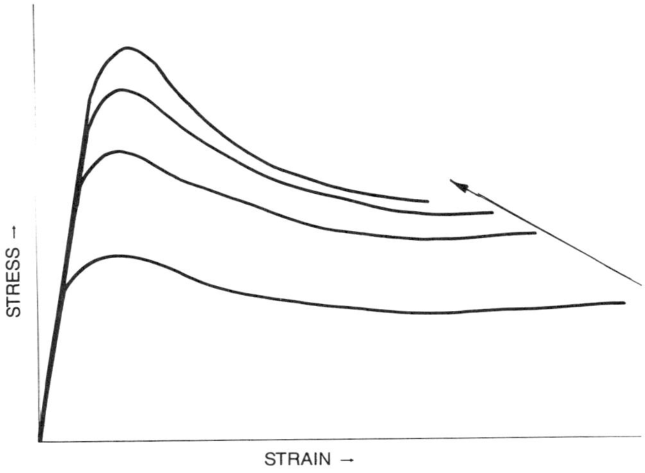

Figure 4.9. THE EFFECT OF STRAIN RATE ON COMPRESSIVE BEHAVIOR

As in Figure 4.8, each curve shows the compression test done at a constant strain rate with the arrow indicating the direction of increasing strain rate. Most polymers show a more definite elastic region in compression than in tension and a comparison of Figures 4.8 and 4.9 will reveal this. It can also be seen that tension testing is more strain-rate sensitive than compression testing in these examples.

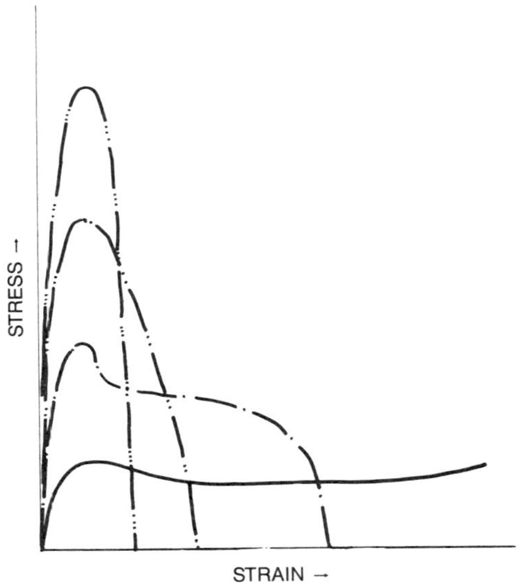

Figure 4.10. THE EFFECT OF PRESSURE ON TENSILE BEHAVIOR

The solid line in this graph represents tensile behavior at atmospheric pressure. As the number of dots within the gap of the dashes increases, so does the applied pressure under which the polymer was tested. It can be seen that as pressure is increased the strength of the polymer is markedly increased with a corresponding decrease in ultimate elongation.

resulting in no noticeable change in the polymer's behavior. This often comes about at times when a very high strain rate is used. The fast testing does not allow the polymer, which has poor thermal conductivity, to dissipate the internal energy it has stored, with the result that the material heats up during testing. A general graph of the effects of strain rate on tensile behavior can be seen in Figure 4.8, and the same type of curve for compression can be seen in Figure 4.9.

THE EFFECT OF HYDROSTATIC PRESSURE ON TENSILE BEHAVIOR

As HYDROSTATIC PRESSURE is increased, the strength of a polymer invariably increases. This is as would be expected since increasing pressure decreases free volume and increases the strength of the van der Waals bonds along with the DENSITY of the material. What would not be expected, however, is the myriad of different responses with respect to ELONGATION and TOUGHNESS. Polymers which are brittle under normal testing conditions may undergo a brittle-ductile transition as a function of increasing pressure. Ductile polymers usually show the reverse trend and become more brittle with increasing pressure. Still another response to the increase in pressure is an initial decrease in toughness followed by an increase in toughness. The graph of the tensile behavior which shows brittle to ductile behavior is similar to that of Figure 4.5 where a decrease in temperature will have the same effect as an increase in pressure. A graph of increasing brittleness with increasing pressure can be seen in Figure 4.10. Figure 4.11 shows an initial decrease in toughness followed by an increase in toughness which is characteristic of some ductile polymers.

ORIENTATION INDUCEMENT DUE TO TENSILE TESTING

As the polymer is stretched during tensile testing it responds to this in a variety of ways. The crystalline region of the polymer is better understood than the amorphous region since the crystalline region responds to structural change analysis methods more directly than does the amorphous region. Thus, the crystalline response will be considered in depth and separately from the amorphous region in the sections that follow. What is perceived of the amorphous response will be contained in this section.

Amorphous polymers respond to an applied force in two ways. The bonds of the backbone can elongate in the direction of the force and the entire chain can straighten. Which happens first is a function of the relative force of the bonding strength of the backbone atoms versus the mobility of the chain segment to rotate into alignment. Elongation of the bonds is a completely elastic process under all conditions. A schematic of normal bonds versus elongated bonds can be seen in Figure 4.12. The alignment of a chain by segment rotation may or may not be elastic depending on the amount of relative motion that is required for alignment. If a chain has to move a great deal relative to its neighboring chains then it will not reassume its original position relative to its neighbors once the force has been removed. It will not have returned to its original position so it will have, by definition, undergone plastic deformation. Alignment of the chains in an amorphous polymer can be taken to

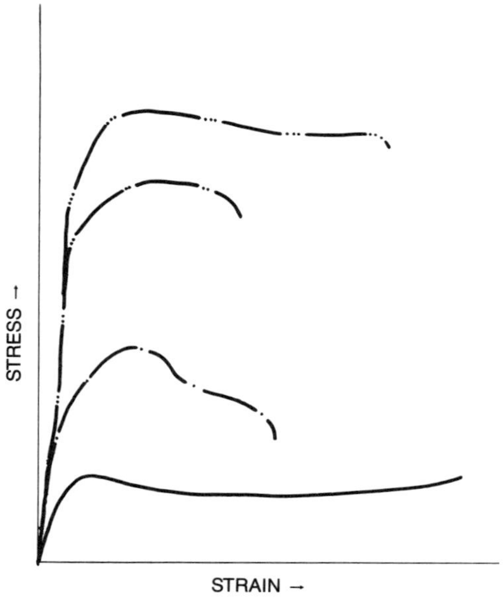

Figure 4.11. OTHER POSSIBLE EFFECTS OF PRESSURE ON TENSILE BEHAVIOR
As in Figure 4.10, this graph shows a tensile test done at atmospheric pressure with a solid line. Also, as the number of dots within the gap of the dashes increases, so does the applied pressure under which testing was done. The same marked increase in strength with an increase in pressure can be seen, but this time the elongation response is not as consistent. What is shown here corresponds roughly to a ductile-brittle-ductile behavior.

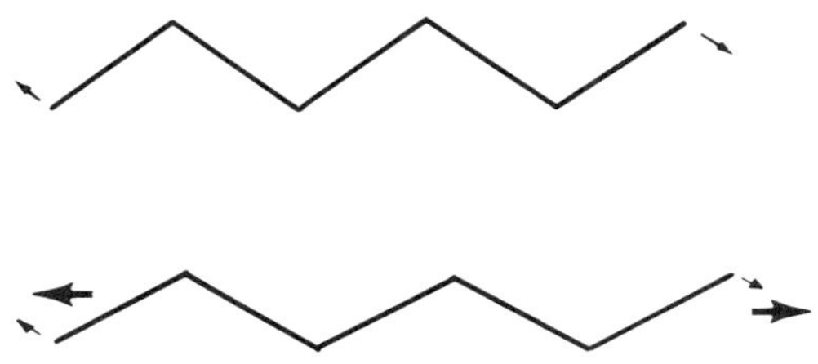

Figure 4.12. ELONGATED BONDS DUE TO AN APPLIED FORCE
The upper schematic is that of a section of polyethylene under normal conditions. The small arrows at each end designate the continuation of the chain. The lower schematic is the same section of polyethylene with an applied force, which is shown by the larger arrows. The difference between the two schematics is the enlargement of the bond angle from 110° to 120°. Although this creates only about 7% elongation in the lower section, it has stored a great deal of energy in the distorted bonds.

such a degree that the chains are in a position to crystallize. Natural rubber is an example of a material that can easily crystallize on alignment of its chains. The stretching of segments into alignment can be seen schematically in Figure 4.13.

Once an amorphous polymer has undergone alignment of the chains in the direction of the applied force during plastic deformation, that direction will have different tensile properties than the polymer has in other directions. More of the force is carried down the backbone of the chain in the aligned direction. Hence, this direction shows a higher strength than the other directions, which have a larger dependence on the van der Waals forces between chains.

TENSILE PROPERTIES OF CRYSTALLINE POLYMERS

The introduction of crystalline structure into a polymer causes changes which directly impact its tensile properties. The crystalline segments act as rigid areas in comparison to the amorphous regions. As such, they can act like additives in increasing the strength of the polymer. With this increased strength usually comes increased brittleness and decreased elongation to failure. If the crystalline phase is more dense than the amorphous phase, then the introduction of the crystalline phase will also increase the initial elastic modulus and the ultimate strength of the polymer, since these properties depend directly on the density of the material.

In a manner analogous to that discussed for additives, if a crystalline region does not move easily with the amorphous region on elongation, that crystalline region can act as a stress raiser and thereby weaken the polymer. This is often the case when the amorphous material is brittle and of very different density than the crystalline component. When the crystalline region does weaken the material, this effect can be

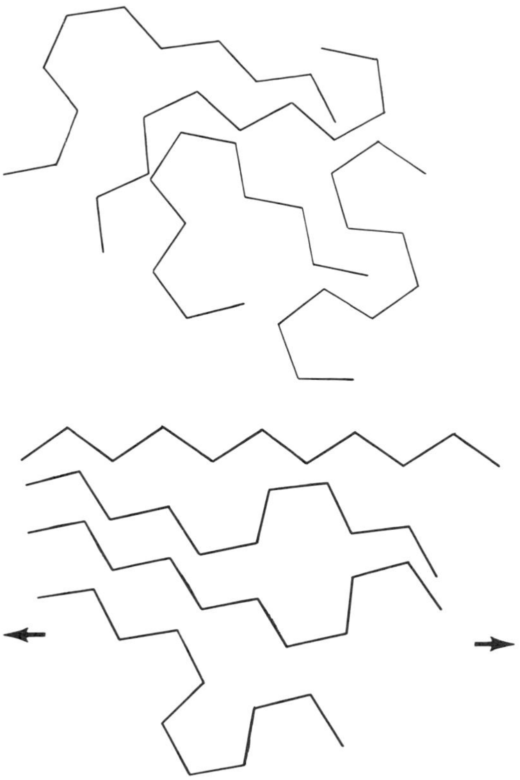

Figure 4.13. ALIGNMENT OF CHAINS DUE TO APPLIED FORCE
The upper schematic shows segments of polymeric chains without an induced alignment. The lower schematic shows these same segments after a force has been applied. The direction of the force application is indicated by the arrows. In both cases four chains of ten segments each are shown with their normal angular relationship. This type of response to an applied force, the uncoiling of a chain wihtout bond distortion, causes a great deal of elongation relative to the amount of force applied but has less ability to store energy than bond elongation.

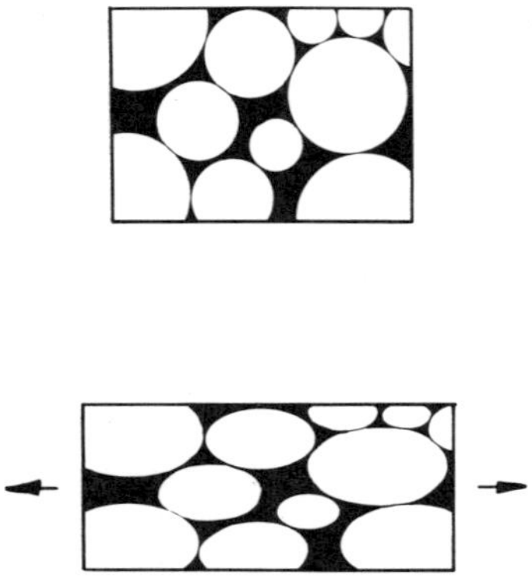

Figure 4.14. ELASTIC DEFORMATION OF SPHERULITES

The top schematic shows idealized, undeformed spherulites within a polymer. The bottom schematic shows the polymer after deformation. The direction of the applied force is indicated by the arrows. Note that the spherulites have not moved relative to one another, so they can easily revert to their initial condition. This is a highly exaggerated view of elastic deformation of a spherulite since true elastic behavior is for extensions on the order of 0.5%. However, the amount of plastic deformation is still quite small for elongations as much as 100% in some polymers. Thus, a mainly elastic deformation can resemble the bottom schematic.

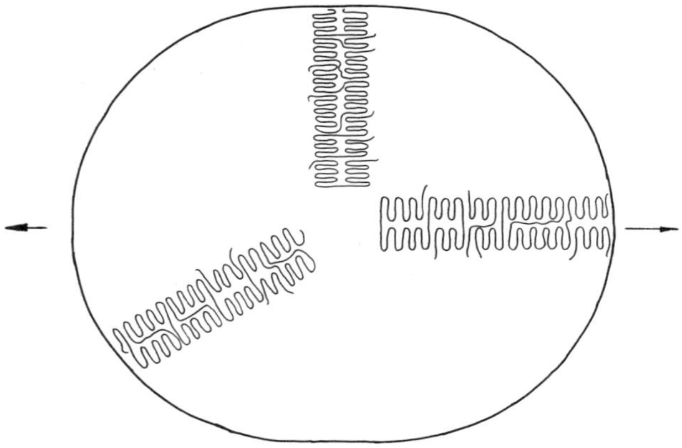

Figure 4.15. RESPONSE OF LAMELLAE DURING ELASTIC DEFORMATION

The oval represents the boundary of an elastically deformed spherulite. Arrows mark the direction of the applied force. Within the oval, several fibrils with associated lamellae have been drawn to illustrate the relative changes that will have occurred to accommodate the change in shape of the spherulite. The lamellae were drawn much larger than they would actually be relative to the spherulite and not to scale within themselves, so that the changes would be more obvious. In the direction perpendicular to the applied force, the vertical, the fibrils are virtually unaffected since there is little compressibility associated with lamellar spacing. However, in the direction of the applied force, the horizontal, the lamellar spacing can be seen to be increased to accommodate spherulite spread in that direction.

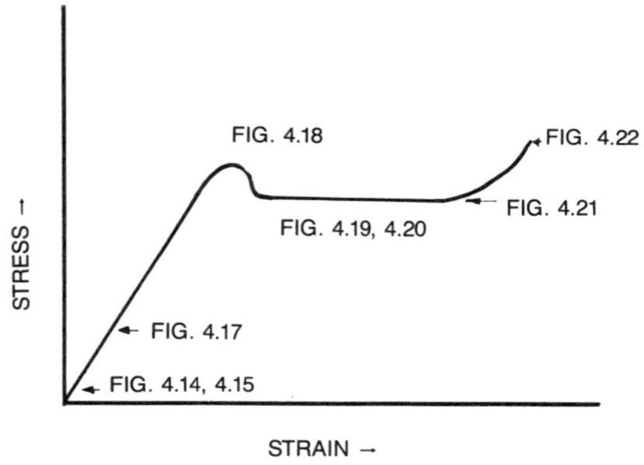

Figure 4.16. THE TENSILE TEST GRAPH OF A CRYSTALLINE POLYMER

This tensile test graph is very similar to the others presented except for the notations along it relative to other figures. These other figures illustrate the approximate structural condition of the crystalline segment of the polymer at that point in the test. This graph can thus be used to correlate the several sections given here on crystalline behavior with an applied force.

minimized or eliminated by orienting the crystalline region relative to the applied stress. In addition, crystalline growth should be encouraged to cut down on the number of possible stress-raising sites. Orientation and crystalline growth of the crystalline portion of a polymer can actually change it from a brittle into a ductile material.

ELASTIC RESPONSE TO TENSILE TESTING OF CRYSTALS

The initial response of a polymeric crystal when a tensile force is applied is to deform elastically. As in the amorphous polymer, the folded chains of the crystal can elongate their atomic bonds to produce a reversible deformation. This has been illustrated in Figure 4.12. The polymeric chain within a lamella cannot "straighten out" as the amorphous polymer can (Figure 4.13), since it is essentially already straight. But it can increase the spacing between the chains elastically. See Figure 1.18 for an illustration of lamellar spacing.

When lamellae are arranged in a spherulite, as they most likely would be, then the spherulite responds as a whole to the applied force. To do this, some lamellae are more affected by the force than others. The lamellae directly in line with the force are the most affected by it, whereas the lamellae perpendicular to the applied force are virtually unaffected. The lamellae in between are elongated by an amount corresponding to their orientation in a uniform way. Thus, the entire spherulite changes uniformly from a sphere to an ellipse with the relative spatial relationships of all lamellae and fibrils maintained. Similarly, the relationship of one spherulite to another will remain constant within the polymer during the elastic phase of deformation. Although this process may well exceed the elastic limit of the amorphous part of the polymer, the overall effect is that of an elastic response of the polymer which is reflected by the initial linear section of the tensile test graph of a crystal-containing polymer. The change in shape of the spherulites due to elastic deformation is illustrated in Figure 4.14. The corresponding effects of the lamellar spacing are shown in Figure 4.15, and the overview of how these physical responses relate to the tensile test graph is shown in Figure 4.16.

PLASTIC RESPONSE TO TENSILE TESTING OF CRYSTALS

Plastic deformation within a crystal is begun when the lamellae move relative to one another. Within a spherulite this will be most evident along the fibrils since the lamellae are well aligned here. The movement of a lamella will be such that it will try to orient its straight chains in the same direction as the applied force. This allows the crystalline structure to carry the maximum load in the direction of the applied force. Thus, the crystal attempts to harden itself in a specific direction when a linear load is applied in that direction. Figure 4.17 shows lamellae within a spherulite responding to an applied force. The onset of slip will begin in different lamellae of a spherulite at different

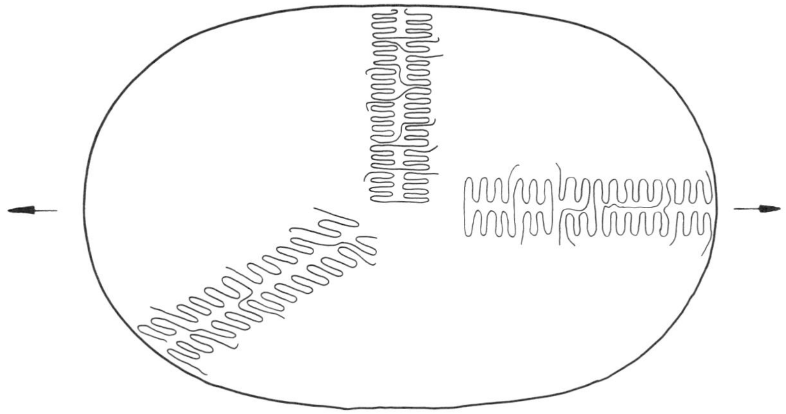

Figure 4.17. PLASTIC SLIP OF LAMELLAE WITHIN A SPHERULITE
This figure is drawn in the same manner as Figure 4.15 and represents the next stage in the response of a spherulite to an applied force. In this stage the lamellae of some of the fibrils are moving relative to one another and thus, plastic deformation has commenced. Under these conditions all of the lamellae will experience some slip except those aligned parallel to the applied force (those on the horizontal). Since these lamellae are already in the alignment of maximum strength relative to the applied force, they have no impetus to change alignment. This anisotropic deformation of the crystalline structure is accommodated by the movement of the amorphous region surrounding the fibrils and lamellae.

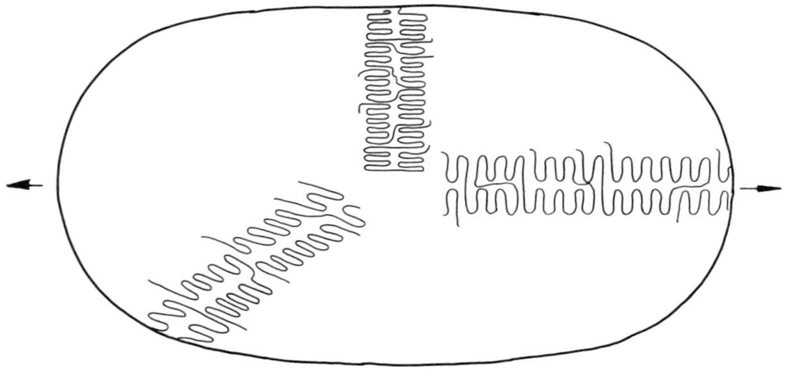

Figure 4.18. TILTING OF LAMELLAE DURING PLASTIC DEFORMATION
This figure uses the same criteria as Figures 4.15 and 4.17. It represents the next stage in the plastic deformation within a spherulite with an applied force, and follows directly from Figure 4.17. At this point the lamellae are experiencing both tilting and slipping within the fibril and the tie molecules are taut. The lamellae have moved as a unit to the extent that they can in most orientations. Of course the fibril parallel to the applied force has remained unchanged since its initial elastic deformation.

times due both to the orientation of the lamellae relative to the applied force and the perfection of the folding.

Once the lamellae have begun to slip relative to one another, they then will tilt so as to further align themselves in the direction of the applied force. The combination of slip and tilt by the lamellae causes a great strain on the tie molecules. Finally, the tie molecules become almost fully extended and cannot accommodate further adjustment of the lamellae as a block within a fibril. Figure 4.18 shows the combination of tilt and slip that some of the lamellae have experienced. As the slip process nears completion, the polymer reaches its yield point as noted in Figure 4.16.

Since the lamellae are unable to move as a block any furhter than they are shown in Figure 4.18, due both to the tension on the tie molecules and the spatial considerations of adjoining lamellae, their next mode of deformation is the unfolding of the lamellae into smaller lamellae with a still closer alignment to the applied force. Figure 4.19 shows this with respect to the spherulite, and Figure 4.20 shows a detail of this figure, the unfolding of a single lamella. The unfolding process continues through the necking portion of the tensile test while the spherulites become ever more elongated, but while they still maintain their spatial relationship to each other.

At the end of necking, the process of reorientation of the lamellae into the direction of applied force is completed, and virtually all lamellae are in their maximum strength position. As a result, the spherulitic structure ceases to exist and the overall appearance of the structure is fibril-like. This new fibril-like structure is not exactly like the fibrilitic structures within

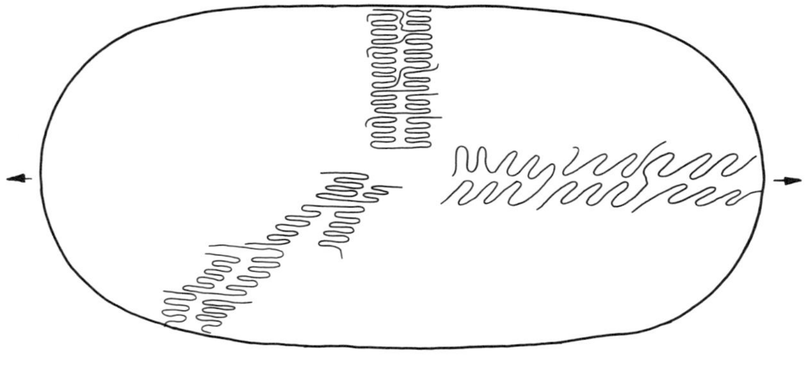

Figure 4.19. UNFOLDING OF LAMELLAE DURING PLASTIC DEFORMATION
This figure uses the same criteria as Figures 4.15, 4.17 and 4.18. It represents the next stage in the plastic deformation within a spherulite with an applied force and follows directly from Figure 4.18. Since the lamellae have both slipped and tilted to reorient themselves relative to the applied force, their next movement in that direction is to unfold and realign in the direction of maximum strength. A larger drawing of a single lamellae unfolding can be seen in Figure 4.20.

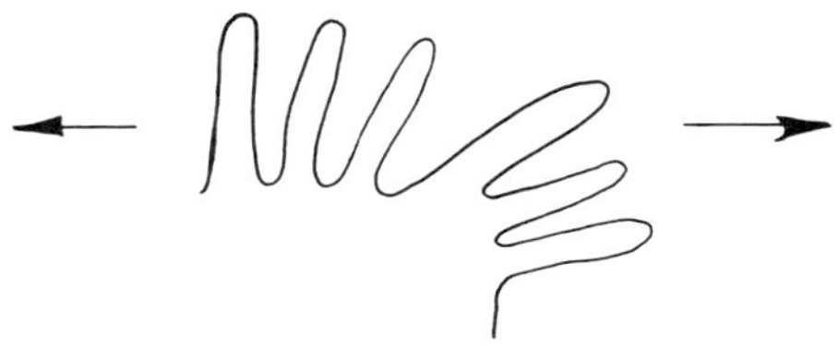

Figure 4.20. UNFOLDING OF A SINGLE LAMELLA
This is a schematic of the side of a single lamella which is unfolding in order to reorient the lamellar folds, so that they can carry the maximum load relative to the applied force, noted by the arrows.

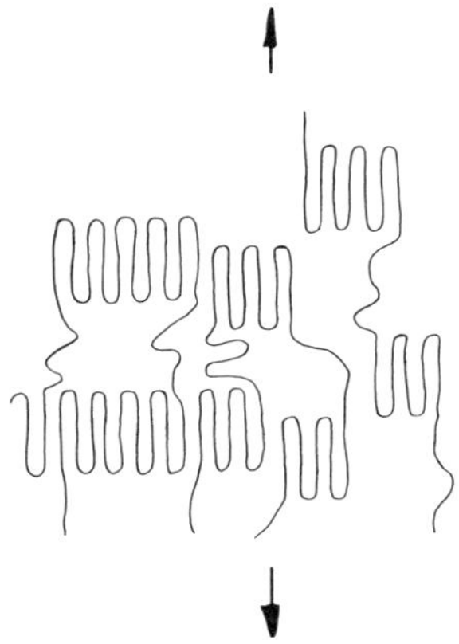

Figure 4.21. CRYSTALLINE STRUCTURE DURING STRAIN HARDENING
This schematic shows several lamellae that have completed reorientation relative to an applied force. The applied force is in the direction of the arrows. The lamellae are now all in the direction of maximum strength relative to the force but there is no overall alignment perpendicular to the applied force. Due to the reorientation, the tension on many of the tie molecules has been reduced and they are no longer at full elongation.

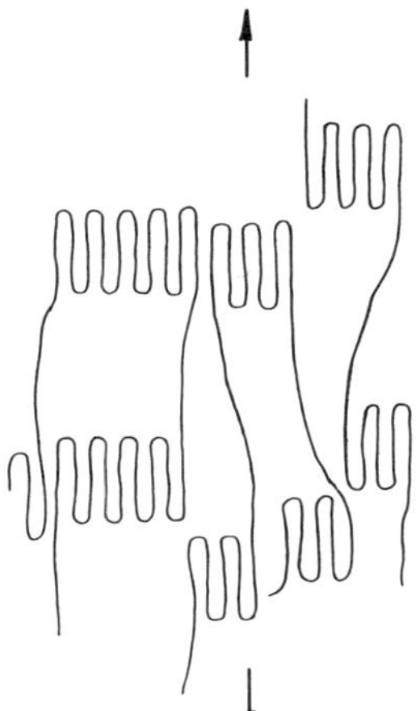

Figure 4.22. STRUCTURE OF FULLY EXTENDED CRYSTAL
This figure is a schematic of a crystalline structure which is at its maximum elongation. The direction of the applied force is noted by the arrows. The lamellae are all in the direction of maximum strength relative to the applied force, and the tie molecules are extended as much as possible. Any further elongation will result in the breaking of many tie molecules and the fracture of the polymer. Under these conditions the tensile strength in the direction of applied force is approximately three times what it is in the direction perpendicular to it. The exact relationship between the strengths in these directions depends on many things; among them are the inter-molecular bonding strength, the stiffness of the backbone and the number of tie molecules. When these factors are optimized, the polymer is a good candidate for use as a fiber.

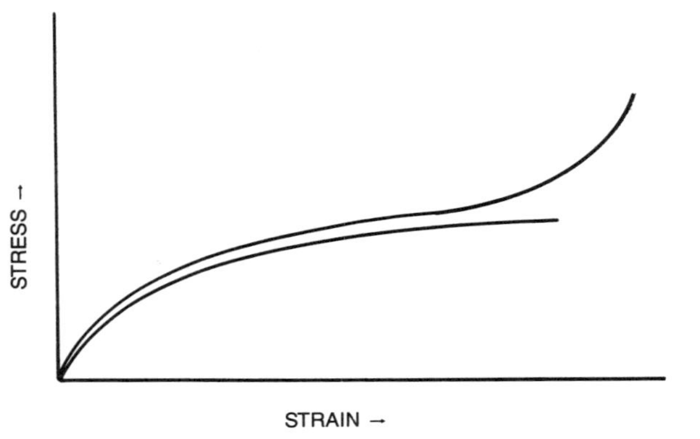

Figure 4.23. TENSILE TEST RESULTS FOR TWO TYPES OF ELASTOMERS

The upper curve is that of a typical elastomer that has formed some crystals during elongation in the tensile test. The initial behavior of this elastomer is very similar to the lower curve which is that of a structurally similar elastomer which does not form crystals on elongation. The first real deviation in these two curves is seen in the middle region where the first crystals and their effects would begin to appear. The fully amorphous elastomer shows consistent strength from the middle region to failure, whereas the crystal-bearing elastomer continues to show increasing strength until failure, especially at the far end of the curve where appreciable STRAIN HARDENING can be seen. The extent and position on the curve of the appearance of strain hardening is highly rate-dependent. If the test is done slowly, the crystals have more formation time and the strain hardening appears sooner and is slightly higher in ultimate value.

the spherulite, since they have aligned lamellae in a radial direction. The fibril-like structure has the lamellae aligned in only one direction rather than in two directions like the radial fibril. Figure 4.21 shows the fibril-like structure.

Further testing causes the lamellae to move in the direction of the applied force, to the extent the tie molecules can elongate. At the maximum elongation, the tie molecules finally break causing fracture of the tensile specimen. This last structure is not only characteristic of the final stages of crystal elongation, but also the type of structure which is strived for in the making of polymeric filaments for yarns and ropes. See Figure 4.22 for a schematic.

TENSILE PROPERTIES OF ELASTOMERS

Elastomers are normally amorphous polymers above the glass transition temperature. They exhibit exceptional ability to elongate under an applied force and subsequently return to their original shape. They are not entirely elastic in their response but can generally be considered so. What distinguishes an elastomer from other polymers in the rubber state is the ability of the elastomer chains to undergo extensive straightening without moving relative to neighboring chains. This ability comes about both by the flexible nature of the backbone of an elastomer and cross-linking that prevents extensive slipping. Thus, on stretching, an elastomer is able to straighten its chain to the extent that elongations up to 1000% are obtainable without the chains moving relative to each other, since the cross-linking requires the chains to deform as a unit. As may be imagined, polymers with flexible backbones would not require a great deal of applied force to straighten. In fact, amorphous elastomers have very low strengths relative to the average polymer in the rubber state.

The process of elongation greatly enhances the alignment of the chains within the elastomer and allows for crystallization on a local basis within some elastomers. The crystalline form of an elastomer is both stronger than the amorphous phase and better able to move with it perfectly during elongation. Thus, the crystals are able to strengthen the elastomer in a way similar to a reinforcing additive. This reinforcement by crystal formation is only in effect during elongation of the elastomer, since removal of the applied force allows the elastomer to revert to its former fully amorphous structure.

The amount of crystalline formation possible and the time required for it to form are subject to the kinetics of crystal formation. Temperatures above the glass transition always allow some crystal formation within an elastomer which is crystallizable. The closer to the glass transition temperature the elastomer is, the more crystalline nucleation sites can exist. However, the lower the temperature, the slower the crystalline growth. Usually, growth of the crystal would need to be very rapid if the usefulness of this mechanism is to be realized. As such, the balancing of these two factors of crystal nucleation and crystal growth results in a maximum effectiveness of this mechanism during tensile testing at about the middle temperature of the rubber state. Optimizing of crystal formation can reinforce the elastomer to such an extent that its strength can be three times that of a completely amorphous elastomer. Figure 4.23 shows the tensile test results for a typical fully amorphous elastomer and for a crystal-forming elastomer.

The amount of cross-linking in an elastomer also affects the tensile properties. Without any cross-linking, the polymer is simply a soft material capable of moderate elongation before failure. Some cross-linking increases the elongation which a polymer can sustain, so that it may then be classified by the general term *elastomer.* While the cross-linking may easily double the possible elongation of the polymer, it will not contribute significantly to its strength. Additional cross-linking will continue to improve the strength of the elastomer but will quickly reach a point where the ultimate elongation will be reduced with each corresponding increase in cross-linking.

The cause of the increase in strength with increasing cross-linking is a combination of two factors. Cross-linking sites will not be as mobile as the rest of the backbone, so they will tend to stiffen the backbone and thus add to the strength of the polymer. The addition of cross-linking sites also adds to the overall rigidity of the material by producing an ever more restricted structure, which is bonded in three dimensions. In the extreme, the material will have so many cross-linked sites that it will be rigid both with respect to backbone motion and relative motion of the chains and will behave more like a loose ceramic crystalline material than the two-dimensionally bonded chains of a polymer.

The correlation between amount of cross-linking and ultimate elongation of the polymer is not as straight-forward as the relationship between cross-linking and strength. The first bit of cross-linking dramatically increases elongation since the structure of the overall material goes from that of individual chains linked only with weak van der Waals forces to chains bound by a few (but strong) covalent bonding sites. Thus, the polymer becomes a quasi-single molecule almost immediately. Additional cross-linking continues to strengthen the response of the material as a unit and thus allows for retention of the relative relationships of the chain segments for elastic deformation. However, a point is reached at which the benefits of the cross-linking sites for unity of motion of the chain segments begin to be outweighed by the ever continuing decrease in the ability of the material to deform under an applied load. Each cross-linkage site acts as a rigid node and tends to restrict the motion of the chain segments adjacent to it. When the cross-linking sites finally become so numerous that the number of atoms of the chain between sites is about equal to the number of atoms in a chain segment, then all elongation of the structure ceases. With only the very limited elongation of the covalent bonds available for response to an applied force, the material is only slightly elastic. In addition, the structure has become so rigid that plastic flow is also impossible, and the material behaves in a brittle manner. See Figure 4.1 for an example.

Mechanical Properties 5

CREEP

All polymers deform plastically over time due to creep. Such creep deformation may be insignificant, as in the case of a polymer below its glass transition temperature and only supporting its own weight, or great enough to cause failure. The factors that are involved in creep are those that operate during a tensile test with the added aspect of relaxation time.

Most creep testing is done at relatively high loads so that the test can be concluded in an expedient manner and the results more easily evaluated. The actual magnitude of a "high" load is determined via a tensile test conducted under conditions similar to that of a creep test. This high load creep information is directly applicable to creep evaluation of a polymer if it is to be used under these test conditions. However, it is difficult to extrapolate such data to other loads since the relationship between load and creep response is usually non-linear.

When a polymer is initially loaded for a creep test it experiences both elastic deformation and plastic flow. It is only the plastic aspect that is considered to be the permanent deformation due to creep. This, if the load is removed, the polymer will have undergone a measurable and predictable elongation due to creep. If this same specimen is immediately reloaded and creep allowed to proceed, and the specimen once again is unloaded and measured for deformation, the second elongation will be greater than the first, even if the test conditions are the same. This is due to the failure of the polymer to relax, or return to its normal structural and stress distribution. However, if the specimen is given sufficient relaxation time between loading, then the elongations due to creep from the first and second loading will be the same if the test conditions are identical. It is this relaxation time factor that makes creep non-linear. Thus, for conditions in which the relaxation time is very fast or sufficient time is allowed during testing, elongation is a linear function of time and creep is linear, otherwise it is not. Figure 5.1 shows creep response during loading and unloading of a typical polymer.

In general, polymers have slow relaxation times and a tensile test is normally completed before the polymer will have relaxed. However, creep testing can be done slowly enough to pemit the polymer to respond by relaxing the localized stresses induced by the load. Relaxation is believed to occur by segment motion. Relaxation times are very long at low temperature where the segment motion is sluggish but can be quite short, approaching that of a metal, at high temperatures.

Relaxation time for a polymer is not easily established and so a calculable variation of creep deviation from linearity with conditions is not usually possible. However, there are some trends in relaxation time which can help one to anticipate the variations from linearity in creep response. Relaxation time is long for all polymers at low temperatures and below the glass transition temperature. The addition of cross-linking also slows relaxation, as does the exceptionally high molecular weight of a polymer. The primary effect of crystallinity is to lengthen relaxation time, although this is often complicated by crystal growth, recrystallization and reorientation under an applied stress. In general, a long relaxation time indicates slow stress dissipation, higher internal energy, a lower creep rate and a higher incidence of brittle failure. In the case of a short relaxation time, the reverse of the above is usually true.

Very complex relationships between structure, temperature and relaxation time occur during a state change. As an example, the glass transition temperature is characterized by a maximum in the creep rate. This has been partially ascribed to the sudden increase in free volume which accompanies the transition. This additional space for movement allows a decrease in the relaxation time.

Under normal applications, an amorphous polymer below its glass transition temperature will only display a maximum elongation due to creep of about 2%

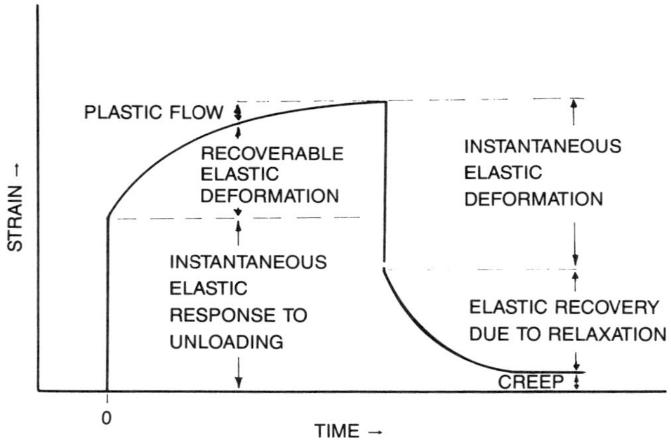

Figure 5.1. CREEP RESPONSE OF A TYPICAL POLYMER
This graph shows the creep behavior of a typical polymer which is initially loaded for testing and then unloaded. The time-axis zero-point has been offset from the strain axis to clarify the initial loading response. When the polymer is initially loaded it responds by deforming elastically. Since the loading is usually faster than the relaxation time of the polymer, it also undergoes some non-linear elastic deformation. Finally, plastic flow begins and continues for the duration of the test. If the polymer is unloaded before failure, the initial elastic response is immediately recovered. With time, the non-linear elastic deformation is also recovered, leaving only the permanent deformation due to plastic flow.

before failure. The creep rate variation is not significant below the glass transition temperature. The creep rate then goes through a maximum during glass transition and stabilizes to a moderately increasing function of temperature in the rubber state. After initial loading response, the creep rate at any specific temperature is usually a constant.

The introduction of a crystalline portion into the amorphous polymer usually has a strengthening effect, and the creep rate and elongation are both reduced. However, the crystalline reaction to an applied load is to reorient the lamellae, creating directionality in the creep response and non-linearity in the creep rate. Recrystallization sufficiently complicates the creep response, such that specific experiments under the desired conditions are prudent for design use if recrystallization is anticipated.

FATIGUE

The cyclic loading and unloading of a polymer induces a fatigue response that is very complex. As with tensile testing (or compression testing), the load that is to be applied, the speed of application of that load, the relative size of the sample, and the temperature all greatly affect the results. An additional factor in fatigue testing is the repetition of the loading. The polymer may be completely unloaded between cycles or only partially unloaded. There may be a long time between cycles or the polymer can be instantaneously reloaded. All of these factors will affect the number of cycles which a polymer can endure before failure. Thus, the usual fatigue data which are given as a function of maximum stress to number of cycles to failure can only serve as a starting point for most applications. Figure 5.2 shows a fatigue graph for a typical polymer.

A factor which greatly alters the expected lifetime of a polymer in fatigue is the unloading extent and duration. If the polymer is completely unloaded between cycles and allowed to remain so for the length of its relaxation time, then it can recover from the loading cycle. Recovery allows the segments of the polymer to attain more favorable alignment for the next loading cycle and thus reduces the plastic flow it will experience. Consequently, the plastic deformation per cycle will be less. In addition, this extra time will allow whatever heat was generated by loading to dissipate. Even if recovery is not possible after each load

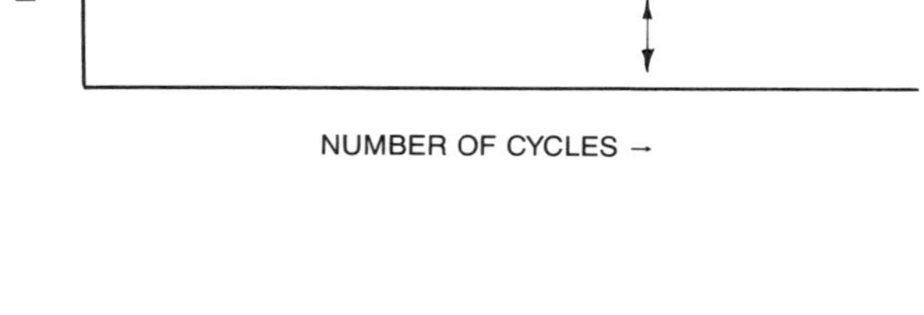

Figure 5.2. FATIGUE CURVE FOR A TYPICAL POLYMER
This graph shows maximum stress plotted versus number of cycles to failure for a typical polymer being tested for fatigue lifetime. It is clear from the shape of the curve that the number of cycles to failure continues to increase with a corresponding decrease in maximum stress until a limit is reached. This limit, the endurance limit designated by the arrow, is sometimes listed in tabular data. It is a practical limit to fatigue failure that does not consider vibrational energy, which can lead to failure in real polymers.

cycle, a periodic recovery time has been shown to be effective in prolonging polymer service life. Conversely, immediate reloading coupled with incomplete unloading will reduce the expected number of cycles a polymer can withstand. An exception to this is rubber, which crystallizes on loading. The crystals both strengthen the rubber and inhibit the propagation of cracks through it. Thus, a minimum load to maintain the crystalline structure ought to be maintained at all times with this type of rubber.

A polymer will respond to the speed of loading, the absolute load and the temperature of the test in a manner very similar to that expected in a tensile test. These effects are then compounded over many such experiences, magnifying their importance. Obviously, reduced temperature, reduced maximum load and reduced speed of loading will all prolong the fatigue life of the polymer.

Failure during fatigue testing comes about either by crack propagation such that the sample breaks or by thermal build-up such that the sample becomes too soft to continue carrying the designated load. Samples do not usually fail by a combination of these methods since they tend to compensate for each other. Thermal build-up causes segment mobility which blunts the leading edge of the crack and slows its movement. Conversely, cracks are usually surface flaws which help to dissipate excess heat. Both crack propagation and thermal build-up are non-linear functions of cycles experienced by the polymer. Both start out slowly and increase, such that only a very short time/number of cycles separates moderate conditions from failure.

Due to the nature of fatigue failure, sample size and shape affect the expected fatigue lifetime of a polymer. Small, flat samples with large relative surface area are much better at dissipating heat and thus are not as susceptible to thermal failure. Conversely, this type of sample is much more likely to have a flaw that will lead to crack propagation. Thus, shape and size of the part under consideration should be viewed relative to which form of failure is anticipated.

MECHANICAL DAMPING

When a polymer is stressed, some of the energy put into it is stored in elastic and plastic deformation and some of it is converted into heat. It is this heat that is measured as mechanical damping. The amount of damping a polymer displays is related to its structure. Throughout the glassy state, damping is low. Once the polymer transforms into the rubber state, the damping remains low but not as low as it was in the glassy state. However, during the transition the damping peaks. Any other primary or secondary transitions will also lead to peaks in the damping, with the glass transition usually being the largest contributor. A graph of damping versus temperature for a typical polymer is given in Figure 5.3, the shear modulus is also included since it is usually measured as part of the same test.

High damping can be a very desirable attribute in polymer applications in which vibrations are to be reduced or a greater coefficient of friction is needed. Since the transition temperature range of a polymer is so limited, several methods have been devised to

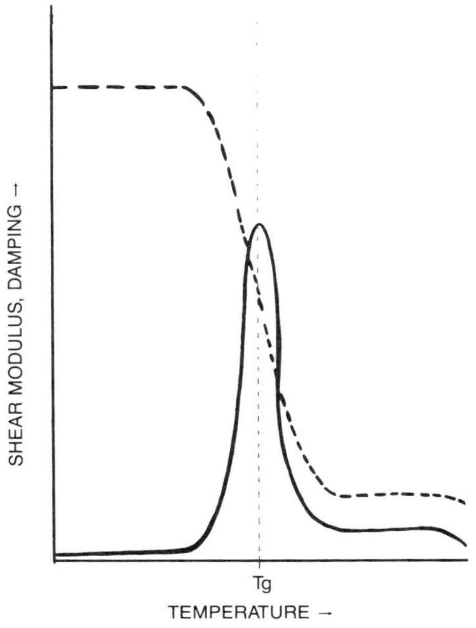

Figure 5.3. MECHANICAL DAMPING OF A TYPICAL POLYMER
This graph shows the general behavior of an amorphous polymer with respect to mechanical damping and shear modulus. The damping is shown by the solid line, and the shear modulus is shown by the dashed line. The vertical dotted line represents the glass transition temperature. The usual spread of the shear modulus and the mechanical damping are both about three orders of magnitude.

broaden it. Plasticizers are an inexpensive and predictable way to broaden the transition temperature range by about 20°C. Poly-blends and copolymers are more effective at broadening the range, but the extent and change they cause in other properties is not as straightforwardly predictable. The complete elimination of transition temperatures by extensive cross-linking has also been used. The resulting polymer usually displays a very consistent, although low, damping over its entire temperature range.

IMPACT STRENGTH

The impact strength of a polymer can be measured in several different ways. The most common is by the Izod method, a pendulum impact, with an unnotched sample. Other means of testing are not directly comparable to this test. Even when the same testing technique is used, the results may vary considerably since the test is sensitive to sample thickness, surface roughness and surface flaws. Using a notched sample for testing helps to limit the surface effects but adds a stress concentration at the tip of the notch that may not simulate service conditions.

Besides those data available on impact strength, several polymeric characteristics can be used to deduce expected impact strength. Since the impact strength is dependent on the ability of the polymer to absorb energy very rapidly, a structure that is more free to move will respond better to the impact. Thus, the rubber state will have a higher impact strength than the glassy state and the impact strength will increase with increasing temperature. Cross-linking will reduce mobility, so the greater the cross-linking the lower the impact strength. Highly cross-linked polymers will not be sensitive to temperature increases with respect to impact strength because the increased mobility due to temperature will not sufficiently alter the overall mobility of the superstructure.

Additions to the polymer can aid or hinder the impact strength depending on their relative impact resistance and mobility within the polymer. Brittle additions such as glass fibers will reduce the impact strength in an already brittle polymer. However, if the polymer is sufficiently mobile to allow flow around the additive, then the stresses will not be concentrated enough to cause localized failure, and the polymer can transfer some of the force of impact to the additive. Thus, an additive may be detrimental to the impact strength of a polymer when the polymer is in its glassy state but advantageous to it when it's in its rubber state.

This same effect is seen with an increase in crystallinity in a polymer. The crystals are usually much more brittle than the amorphous polymer and act as stress concentrations. This is exacerbated if the crystals are in the form of large spherulites or if the crystallinity is very high. Thus, the crystals will decrease the impact strength when the amorphous part is brittle. This effect is reversed when the amorphous part is flexible and the crystal size and number are small; then the crystals act as reinforcing additives.

When an additive is itself high in impact strength, the overall impact strength of the polymer to which it is added will generally be improved. The amount of this improvement depends on several factors. Small size and relatively low concentration of the additive are generally considered preferable in both the effectiveness of the addition and the minimizing of the adverse effects of an additive.

FRICTION

Like most mechanical tests performed on polymers, those used to determine the coefficient of friction produce results that vary considerably. Much of this variation is due to testing conditions. Polymers are very sensitive to the testing surface, the rate of movement, the size and shape of the sample, and the number of passes made.

When a polymer is tested against a surface harder than itself, the results will be much more consistent and reflective of the coefficient of friction of the polymer than those obtained against a softer surface or against itself. A harder surface will tend to minimize the effects of adhesion between the polymer and the testing surface. In addition, the testing surface should not have a bonding affinity for the polymer which will promote adhesion. A polymer tested against itself may have a higher coefficient of friction than against a softer surface because of bonding attraction producing high adhesion. Similarly, polymers with polar dipoles will have a higher attraction and adhesion to metal testing surfaces than symmetrically bonded polymers. However, if the dipoles are able to attract water to the surface of the polymer, then the water can act as a lubricant, reducing the coefficient of friction.

The coefficient of friction tends to decrease with an increase in rate of testing until a minimum is reached; beyond the minimum, the coefficient of friction increases with rate of testing. The minimum comes about when a number of factors are in balance. From the mechanical point of view, friction would always decrease with increasing rate; however, heat is generated which usually limits this. Heat is less easily dissipated by the polymer than other materials so it tends to have a greater effect. In general, an increase in internal temperature of a polymeric sample, or an increase in the temperature of the testing conditions will increase the coefficient of friction. A significant exception to this is PTFE, which shows no real temperature dependence of the coefficient of friction over the range from 20° to 200°C.

The size and shape of a sample can alter the results of friction testing because of the nature of the polymer interface with the testing surface. Under most testing conditions, the polymer will be subject to both elastic and plastic deformation. The plastic deformation will create alignment of the chains in the polymer relative to the applied force, and alter the surface. The plastic flow will also tend to increase the contact area of the sample with the surface and thus increase the measured coefficient of friction. One means of preventing significant surface contact between polymer and testing surface is to coat a harder core material with the polymer. This way only a thin film of the polymer would be tested and significant deformation cannot occur. If the core is metal, it will also dissipate heat created during testing, further helping to maintain dimensional stability as well as control the rise in temperature. The combination of a metal core and a polymeric film has found many useful applications where low coefficient of friction is desired. However, the absence of a polymeric mass does not allow for the absorption of vibration that is usually considered desirable in a polymeric part.

During repeated passes of a friction or wear sample, both contacting surfaces are subject to mass transfer. For slow speeds and low applied forces mass transfer will be slight. Above moderate limits, however, such transfer can greatly increase the measured coefficient of friction. Eventually the surface of the polymer picks up an equilibrium amount of the material that comprises the other testing surface and vice versa. At this point, the measured coefficient of friction should remain consistent and will be generally higher than the values first recorded.

FRACTURE

The normal mode of failure of a polymer under an applied tensile load is fracture. Fracture is a combination of the two failure mechanisms available to a polymer, slippage of the chains and chain scission. For most polymers the amount of slippage that the chains can support is a small part of the fracture damage, even when the loads are low and the time to fracture is long. The immediate observable effect of this mode of failure is stress whitening. The slippage that does occur increases the localized stress level and encourages chain scission. Chain scission during fracture comes about when the stress on a small segment of the chain is greater than the bond strength can sustain, and one or more primary backbone bonds break.

The initiation of fracture can be safely assumed to occur at a flaw within the polymer. The region around a flaw, such as a void, will not be able to sustain the applied force along with the unflawed regions of the polymer, and crazing will begin, followed by cracking. Once cracking has begun it is only a matter of time until it propagates. The time can be quite slow if the applied force is low and the polymer is in the rubber state. It can be nearly instantaneous if the applied force is near the limit of strength of the polymer and the polymer is in the glassy state.

The rapidity of the application of the applied load also affects the failure rate of the polymer. If a polymer with an existing crack experiences an impact, the energy required to produce failure is very much less than if the crack had to be formed or nucleated. The state of a polymer will also affect its response to impact. Glassy state polymers have long relaxation times and are not as free to absorb the energy of the impact by chain slippage, so almost all of the impact energy must be absorbed by the chain backbones. If failure occurs, it will be by instantaneous brittle fracture.

Once a craze begins to form at a flaw, the polymer chains in that region reorient in response to both the applied load and the reduced cross-sectional area or stress concentration at that point due to the flaw. That is, the chains align in the direction of the applied force to better carry the load. Figure 5.4 shows schematically the unaligned chains in the region of a void, and

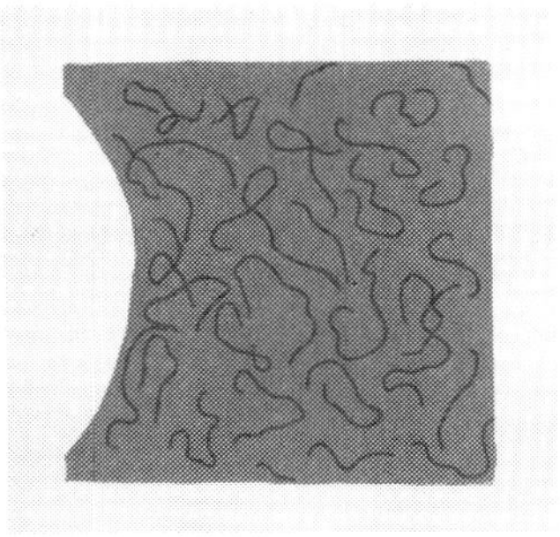

Figure 5.4. THE STRUCTURE OF A VOID
This schematic shows a void within a polymer. The light gray area is that of the bulk of the polymer with some black lines drawn in to represent a few specific chains. Note that the void has a rather smooth surface with no localized stress concentrator.

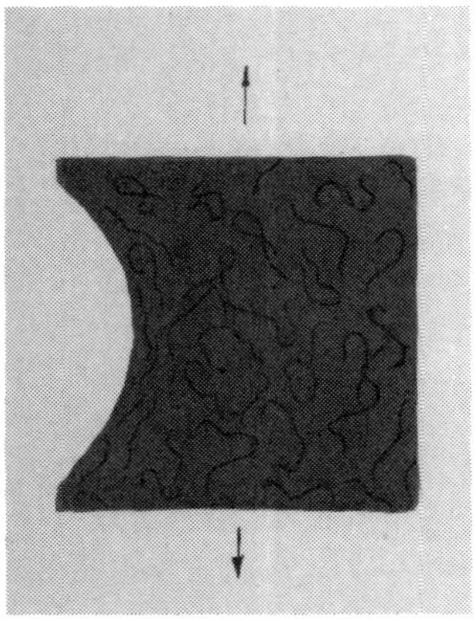

Figure 5.5. THE ONSET OF CRAZING AT A VOID

This figure represents the same area as Figure 5.4 after a tensile load has been applied to the polymer. The load is indicated by the arrows. The chains nearest the void experience the highest localized force since the void does not contain load-bearing material. These chains align in the direction of the applied force to better carry the load.

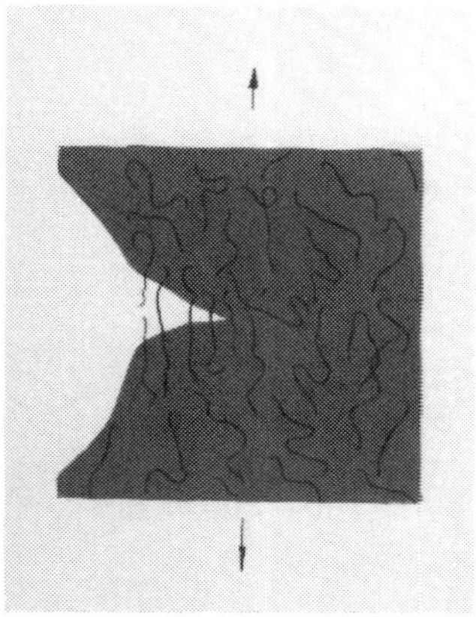

Figure 5.6. CRACK INITIATION WITHIN A CRAZE

This figure also represents the region shown in Figure 5.4. After the craze was formed by the applied load, a crack initiated due to the most stretched chain's inability to further respond to the loading force. As each chain is stretched to its limit and breaks, it places that much more burden on the adjacent chains, which stretch and break in turn through the polymer until fracture is complete. The applied load is indicated by the arrows.

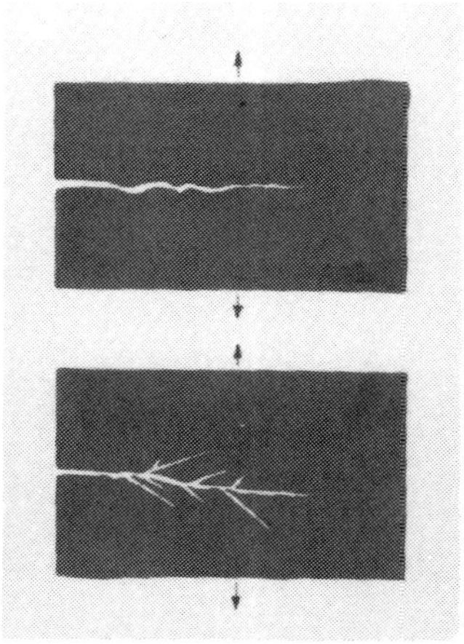

Figure 5.7. CRAZING AND BRANCHING DURING FRACTURE

The top figure shows how a craze would appear that has propagated perpendicular to the applied force in a polymer. The polymer is depicted in gray and the applied force is shown by the arrows. The bottom figure shows the same polymer when the conditions have been favorable for both crazing and branching. The bottom figure will most likely still break perpendicular to the applied force, but the branches will have extended to damage the polymer beyond the fracture surface.

Figure 5.5 shows that same region after a force has been applied. Once the craze has formed, the volume of polymer in that region able to carry the load is reduced and unable to support chain slippage. Thus, the next step is for the chains to begin breaking at the surface between the flaw, e.g., void, and the craze. As the chains break, there is less and less material available to carry the applied force and the craze continues to advance into the polymer. This then allows for the crack to follow behind the craze. Figure 5.6 shows the same region as Figure 5.4 after some chain breakage.

Crazes and the cracks that follow propagate perpendicularly to the direction of the applied force unless other factors influence their path. Additives are one of the factors that can vary the path of the craze/crack. When an additive that is not as strong as the crazed region around it is encountered the crack may propagate right through it, provided that the additive is "anchored" within the bulk of the polymer. This is often the case for additives in the form of brittle fibers. If the additive is engulfed within the craze it may have no effect on the direction of the propagation. Larger additives and crystalline regions are likely to divert the craze around them unless cracking through them is possible. If the craze is diverted around a structure it may continue on an angle from the original craze or branch. Branching also is displayed in polymers when the craze does not have sufficient time to develop completely. This will happen at both low temperatures and at high speeds of load application. Thus the polymer shatters instead of breaks. Figure 5.7 shows diagrams of crack branching.

Factors that weaken the interchain attraction and increase the mobility of the chain segments increase the polymer's ability to craze in response to an applied force. Plasticizers increase the mobility of the polymer and thus encourage crazing. This is true both of intentional plasticizers, such as those used for processing, and unintentional plasticizers such as absorbed water. Absorbed gases can also serve as plasticizers. They are not very important at normal temperatures, but can be very significant at temperatures well below the glass transition temperature.

Cross-linked rubbers are already in an arrangement throughout their structure, which is reminiscent of the localized crazed structure of a non-cross-linked polymer. As such, they do not develop a true crazed region through which a crack can follow. The whole structure is equally strong or weak as the case may be. Failure of a cross-linked rubber does come primarily from chain breakage but at a location where a concentrated force is applied or a flaw is present. This is called tearing and has no intermediate structure between unfailed and failed material.

Crazes do not usually form when a polymer is deformed in compression or shear although they may appear to form. What actually has formed is analogous to the shear band of a metal. These will have the optical opacity of a craze since they also contain reorientated material but the alignment of the chains in the direction of the applied stress will not be as complete. The stress bands will appear at a reproducible angle from the perpendicular angle expected of a craze. A larger volume of polymer will be associated with the shear bands than with crazing if all conditions are the same except the mode of applied force. As with a craze, stress bands can be removed if the material is allowed sufficient relaxation time. This is usually facilitated by an increase in temperature.

FOAMS

The mechanical properties of foams are dependent on the type of polymer used for production, the density of the foam, the presence of a "skin" on the surface, the size of the cells, the uniformity of the cells, the wall thickness of the cells, the closure of the cells, the gas within the cells, the pressure of the gas within the cells and a few million other possible variables. There are no existing models that include all of the variables, or even several of the variables.

For most applications the property in question can be extrapolated from that property in the fully dense polymer. The usual trend is for the property to decrease with decreasing density of the foam. That is, as the amount of polymer available to perform the function decreases, so does the foam's ability to perform the function. As an example, when the density of a foam is decreased, the tensile strength would be expected to decrease. This variation can be further simplified by considering this relationship to be linear. Usually, it is fairly close to being linear within an order of magnitude decrease in density. Very low density foams are too variable for any generality to be given as to their mechanical behavior.

Tensile strength, tensile modulus, compressive strength, compressive modulus, flexural strength, flexural modulus, shear strength and shear modulus all follow a reasonably linear decrease with decreasing density for most polymers at any given temperature. They also all show an alarmingly rapid deterioration of properties, or a steep slope. Thus, some foam data are required to make an estimate of the expected property of a foam at a specific density.

Extrapolation of data from one temperature to another is very risky. None of these properties appears to vary linearly or even with a consistent trend. One explanation for this has been attributed to the role that gas plays within the foam. At high temperatures the gas will expand and display a greater than atmospheric pressure if the polymer is able to contain the gas. This will clearly add to the strength of the foam. At low temperatures the gas may contract sufficiently that the foam will collapse under its own weight.

Impact strength also follows the trends of the other mechanical properties if it is relative to a foam with low defect structure on its surface. Most foams have a very high defect structure on the surface, especially the open celled ones. However, if the foam has been formed against a surface such that a "skin" of relatively dense polymer has formed, then the foam will probably follow the linear relationship between impact strength and density.

Properties of Selected Polymers 6

GENERAL, Proper Name of Polymer, Common Name, Abbreviation

Note: Data are given in metric units with English unit conversions given parenthetically. Additional information which may be of interest is given following a semi-colon, in whatever units it occurs. All data in this section can be assumed to have been obtained to the specifications listed here.

GLASS TRANSITION TEMPERATURE: −50°C (−58°F) for most rubbers, 100°C (212°F) for most high-strength polymers

HEAT-DISTORTION TEMPERATURE: 100°C (212°F) measured to ASTM D648

MAXIMUM CONTINUOUS SERVICE TEMPERATURE: 85°C (185°F) measured to ASTM D759-48

MINIMUM CONTINUOUS SERVICE TEMPERATURE: −30°C (−22°F) measured to ASTM D759-48

MELTING/DECOMPOSITION TEMPERATURES: 150°C (302°F)

SPECIFIC GRAVITY: 1.2 measured to ASTM D792

SPECIFIC HEAT: 1.3 KJ/kgK

THERMAL CONDUCTIVITY: 0.2 W/mK measured to ASTM C177

COEFFICIENT OF LINEAR EXPANSION: 0.0001/K (0.00006/°F) measured to ASTM D696

HARDNESS: 100 N/mm ball indention or R150 measured to ASTM D785

ELONGATION: 100% measured to ASTM D638, varies from 1% to 1000%

MODULUS OF ELASTICITY: 1 MPa (145 psi) for rubbers; varies from 200 MPa (30,000 psi) to 20,000 MPa (3,000,000 psi) for other forms of polymers, measured to ASTM D747

TENSILE STRENGTH: 50 MPa (7,500 psi) measured to ASTM D-638, 651

IMPACT STRENGTH: most polymers do not break for the unnotched test; they break during the notched test measured to ASTM D256 at about 10 J/mm

FLEXURAL STRENGTH: 100 MPa (15,000 psi) measured to ASTM D790

FLEXURAL MODULUS: 3,000 MPa (440,000 psi)

DEFLECTION TEMPERATURE UNDER FLEXURAL LOAD: 66°C (150°F) under a load of 1.8 MPa (264 psi); 93°C (200°F) under a load of 0.5 MPa (66 psi) measured to ASTM D 648

SHEAR STRENGTH: 30 MPa (4,000 psi) measured to ASTM D732

SHEAR MODULUS: 10,000 MPa (1,500,000 psi) for glass and 1 MPa (145 psi) for rubber states, both are relatively independent of temperature within the state

POISSON'S RATIO: approx. 0.3 for glassy state; approx. 0.4 for crystalline state and approx. 0.5 for rubbery state

COMPRESSIVE STRENGTH: same as tensile, measured to ASTM D695

ABRASION RESISTANCE: good

CREEP: 7 MPa (1000 psi) load has creep modulus of 2,100 MPa (300,000 psi) after 100 hrs., less than 2% in glassy state measured to ASTM D2990

COEFFICIENT OF FRICTION: 0.3

ELECTRICAL RESISTIVITY: 10E16 Ohm/cm for volume measured to ASTM D257; 10E14 Ohm for surface

DIELECTRIC STRENGTH: 40 kV/mm measured to ASTM D149

DIELECTRIC CONSTANT: 3 over the 50 to 10E9 Hz spread, measured to ASTM D-150

ARC RESISTANCE: 10 sec. measured by ASTM D495

DISSIPATION FACTOR: 0.0007 at 60 Hz and 0.01 at 10E6 Hz, measured to ASTM D150

REFRACTIVE INDEX: 1.5 measured to ASTM D542

TRANSMISSION OF LIGHT: usually high

MOISTURE-VAPOR TRANSMISSION: 20 g/mil/100 in-in/

24 hrs, measured to ASTM D96-66

WATER ABSORPTION: 0.2% of volume except nylon which is 1.5%; measured to ASTM D570

GAS TRANSMISSION RATE: 300 cc of oxygen through 1 mil/24 hr/100 in-in/atm, 900 cc of carbon dioxide with same units, measured to ASTM D1434-63

FLAMMABILITY: most polymers can burn, measured to ASTM D1433-58

COMBUSTION PRODUCTS: carbon dioxide, carbon monoxide, hydrogen

FLASH IGNITION TEMPERATURE: 315°C (600°F)

LIMITING OXYGEN FOR SUSTAINED IGNITION: 15%

BURNING RATE: rate varies widely, measured to ASTM D635

CHEMICAL STABILITY; RESISTANCE TO:

ACIDS: usually good although an applied stress can cause substantial attack, measured to ASTM D543-63T

BASES: same as acids

SOLVENTS: good except for chemically similar liquids which can be used as a plasticizer, measured to ASTM D543-63T

OILS AND FATS: excellent except natural rubber, measured to ASTM D722-45

SUNLIGHT: poor since most have bonds broken by ultraviolet light, measured to ASTM D1435-58

BACTERIAL AND FUNGAL ATTACK: pure polymer usually not attacked

TOXICITY: usually a function of processing chemicals rather than the polymer itself

RESPONSE TO IRRADIATION: cross-links until high doses

WEAKNESSES: brittleness, processing difficulties, high cost

MONOMER: –C–C–C–C– chain with small side groups

STRUCTURE: thermoplastic, limited crystallinity

APPLICATIONS: wherever light weight, corrosion resistance and moderate strength are required; tires, oil field and chemical processing equipment, electrical insulation, automotive interiors, O rings and seals, gears, home furnishing, apparel, safety helmets, glazing, rope, flooring, bedding, packing film, toys, insulation, caulking, artificial heart valves, boat hulls, tablewear, buttons, hoses, magnetic tape

EPOXY (EP)

GLASS TRANSITION TEMPERATURE: 120° to 190°C (250° to 375°F)

HEAT-DISTORTION TEMPERATURE: 177°C (350°F)

MAXIMUM CONTINUOUS SERVICE TEMPERATURE: 130°C (266°F)

SPECIFIC GRAVITY: 1.11–1.4

SPECIFIC HEAT: 0.8 KJ/kgK

THERMAL CONDUCTIVITY: 0.88 W/mK

COEFFICIENT OF LINEAR EXPANSION: 0.000011–0.000035/°K (0.000006–0.00002/°F)

HARDNESS: M100 Rockwell

ELONGATION: 4%; 5% for 90 Kg/m³ (5.6 lb/ft³) density foam

TENSILE MODULUS: 2150 MPa; 300,000 psi

TENSILE STRENGTH: 28–90 MPa; 4000–13,000 psi; 1 MPa (150 psi) for 90 Kg/m³ (5.6 lb/ft³) density foam

IMPACT STRENGTH: 8 N/mm for unnotched; 3 N/mm for notched

FLEXURAL STRENGTH: 345 MPa (50,000 psi)

FLEXURAL MODULUS: 17,200 MPa (2,500,000 psi)

DEFLECTION TEMPERATURE UNDER FLEXURAL LOAD: 46–288°C (115–550°F) at 1.8 MPa (264 psi)

COMPRESSIVE STRENGTH: 241 MPa (35,000 psi)

ABRASION RESISTANCE: very good

CREEP: under ambient conditions and an applied stress of 28 MPa (4000 psi) the creep modulus is 10,000 MPa (1,450,000 psi) at 1 hr and 8,600 MPa (1,250,000 psi) at 100 hr

ELECTRICAL RESISTIVITY: 10E14 Ohm/cm for volume; 10E12 Ohm for surface

DIELECTRIC STRENGTH: 30–40 KV/mm; 5 KV/mm for 32 Kg/m³ (2 lb/ft³) density foam

DIELECTRIC CONSTANT: 3.5–5 at 50 and 10E6 Hz; 2 at 10E3–10E9 Hz for 320 Kg/m³ (20 lb/ft³) foam and 1.1 for 16 Kg/m³ (1 lb/ft³) density foam

ARC RESISTANCE: 120 sec.

DISSIPATION FACTOR: 0.001 at 50 Hz, 0.01 at 10E6 Hz; 0.005 at 10E3–10E9 Hz for foam of most densities

REFRACTIVE INDEX: 1.6

TRANSMISSION OF LIGHT: 82–90%, usually clear in color

WATER ABSORPTION: 0.05–0.2% of volume; similar for foams

COMBUSTION PRODUCTS: black smoke

BURNING RATE: self-extinguishing

CHEMICAL STABILITY; RESISTANCE TO

ACIDS: excellent, same for foams

BASES: excellent, same for foams

SOLVENTS: variable, same for foams, excellent for aliphatic hydrocarbons, fair for aromatic hydrocarbons and poor for low-boiling ketones

OILS AND FATS: excellent
SUNLIGHT: good
WEAKNESSES: difficult to mold and expensive

MONOMER:
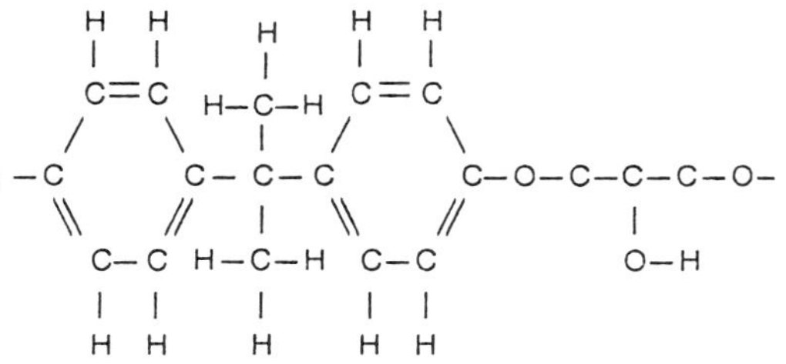

STRUCTURE: cross-linked family of thermosetting polymers which will not crystallize
APPLICATIONS: encapsulation of electrical components due to low shrinkage on curing and excellent electrical insulation, adhesive especially to metal and concrete

POLY(ACRYLONITRILE-BUTADIENE) COPOLYMER, NITRILE (NBR)

GLASS TRANSITION TEMPERATURE: −56°C (−60°F) for 20%/80%; −41°C (−42°F) for 30%/70%
HEAT-DISTORTION TEMPERATURE: approx. 100°C (212°F)
MAXIMUM CONTINUOUS SERVICE TEMPERATURE: 93–121°C (200–250°F)
SPECIFIC GRAVITY: 1.11
ELONGATION: 300–700%
TENSILE STRENGTH:b 7 MPa (1000 psi)
IMPACT STRENGTH: 60 J for notched
FLEXURAL STRENGTH: 120 MPa (17,400 psi)
FLEXURAL MODULUS: 690 MPa (100,000 psi)
ABRASION RESISTANCE: good
ELECTRICAL RESISTIVITY: poor
DIELECTRIC STRENGTH: poor
GAS PERMEABILITY: 4X10E17 cm^3s/g for oxygen, 10E18 cm^3s/g for carbon dioxide, 23X10E17 cm^3s/g for water vapor
CHEMICAL STABILITY; RESISTANCE TO:
ACIDS: good
BASES: fair
SOLVENTS: variable; absorption of kerosene: −5 to 20%, gasoline: −5 to 20%, benzene: 75–250%, acetone: 100–300%, ethyl alcohol: −10 to 16%, carbon tetrachloride: 40–150%, lubricating oil: −3 to 40% increase in volume of NBR
OILS AND FATS: excellent
SUNLIGHT: poor
OXYGEN AND OZONE: poor
TOXICITY: approved for food packaging
WEAKNESSES: reduced resilience and low temperature flexibility over BR

MONOMER:
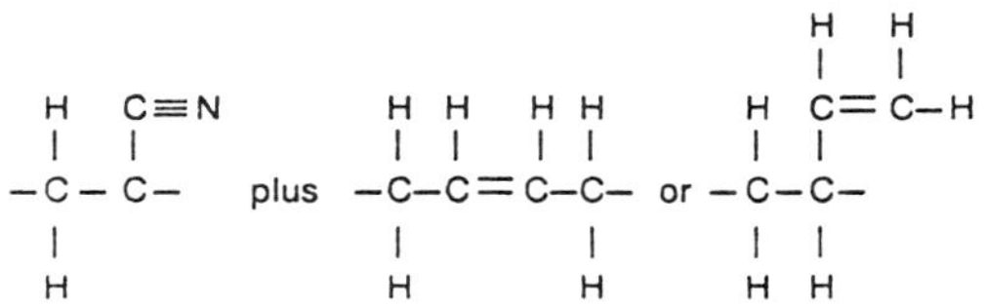

acrylonitrile plus butadiene of one of these two forms

STRUCTURE: copolymer which is vulcanized into an elastomer
APPLICATIONS: O rings and oil seals due to excellent oil resistance; packaging for medicines and household chemicals

POLY(ACRYLONITRILE-BUTADIENE-STYRENE) COPOLYMER (ABS)

GLASS TRANSITION TEMPERATURE: approx. 100°C (212°F)
HEAT-DISTORTION TEMPERATURE: 95° to 100°C (203° to 230°F)
MAXIMUM CONTINUOUS SERVICE TEMPERATURE: 75° to 85°C (167° to 185°F)
MINIMUM CONTINUOUS SERVICE TEMPERATURE: −40°C (−40°F)
SPECIFIC GRAVITY: 1.04–1.06
SPECIFIC HEAT: 1.3 KJ/kgK
THERMAL CONDUCTIVITY: 0.18 W/mK
COEFFICIENT OF LINEAR EXPANSION: 0.00006–0.00011/°K (0.00003–0.00006°F)
HARDNESS: 80–120 ball indention N/mm^2
ELONGATION: 15–30%
TENSILE MODULUS: 1900–2700 MPa; 200,000–500,000 psi
TENSILE STRENGTH: 32–45 MPa; 9000 psi
IMPACT STRENGTH: 70 N/mm to no break for unnotched; 7–20 N/mm for notched
FLEXURAL STRENGTH: 93 MPa (13,500 psi)
FLEXURAL MODULUS: 2,700 MPa; 400,000 psi
DEFLECTION TEMPERATURE UNDER FLEXURAL LOAD: 87°C (188°F) under a load of 1.8 MPa (264 psi); 93°C (200°F) under a load of 0.5 MPa (66 psi)
SHEAR STRENGTH: 28 MPa (4,100 psi)
COMPRESSIVE STRENGTH: 35 MPa (5,100 psi); 69 MPa (10,000 psi)
ABRASION RESISTANCE: good

CREEP: tested with an applied stress of 7 MPa (1000 psi) and ambient conditions, creep modulus is 2,400 MPa (350,000 psi) at 1 hr, 2,100 MPa (300,000 psi) at 100 hr and 1,500 MPa (215,000 psi) at 1000 hr; for an applied stress of 14 MPa (2,000 psi) the creep modulus is 2,300 MPa (340,000 psi) at 1 hr, 2,100 MPa (300,000 psi) at 100 hr and 1,700 MPa (245,000 psi) at 1000 hr

ELECTRICAL RESISTIVITY: 10E14 Ohm/cm for volume; 10E13 Ohm for surface

DIELECTRIC STRENGTH: 35–50 kV/mm

DIELECTRIC CONSTANT: 2.4–5 at 50 Hz; 2.4–3.8 at 10E6 Hz

ARC RESISTANCE: 65 sec

DISSIPATION FACTOR: 0.003–0.008 at 50 Hz; 0.002–0.015 at 10E6 Hz

TRANSMISSION OF LIGHT: 33% with 100% haze; light tan in color and opaque

WATER ABSORPTION: 0.2–0.45% of volume

FLAMMABILITY: burns with green flame

COMBUSTION PRODUCTS: black smoke

BURNING RATE: slow

SELF-IGNITION TEMPERATURE: 466°C (781°F)

CHEMICAL STABILITY; RESISTANCE TO:
- *ACIDS:* excellent except for concentrated oxidizing acids
- *BASES:* excellent
- *SOLVENTS:* poor except amines
- *OILS AND FATS:* excellent
- *SUNLIGHT:* yellows slightly with some embrittlement

TOXICITY: approved for food use

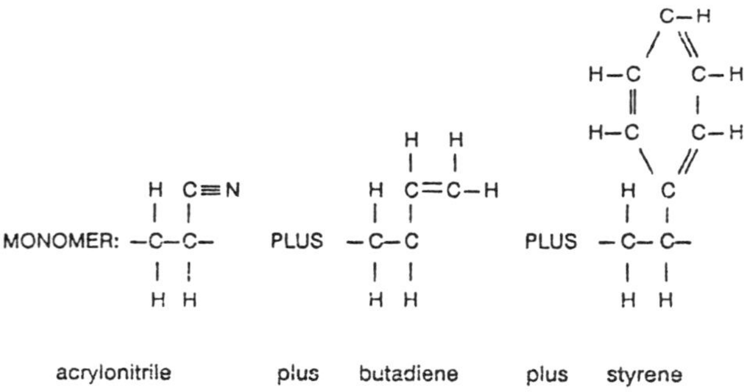

STRUCTURE: graft copolymer which does not crystallize, thermoplastic

APPLICATIONS: dinghy hulls, telephone handsets, housings for vacuum cleaners and lawn mowers and other large forming operations which make use of its dimensional stability and excellent formability

POLYBENZIMIDAZOLE (PBI)

MAXIMUM CONTINUOUS SERVICE TEMPERATURE: approx. 400°C (752°F)

CHEMICAL STABILITY; RESISTANCE TO:
- *ACIDS:* good
- *BASES:* good

WEAKNESSES: relatively high cost

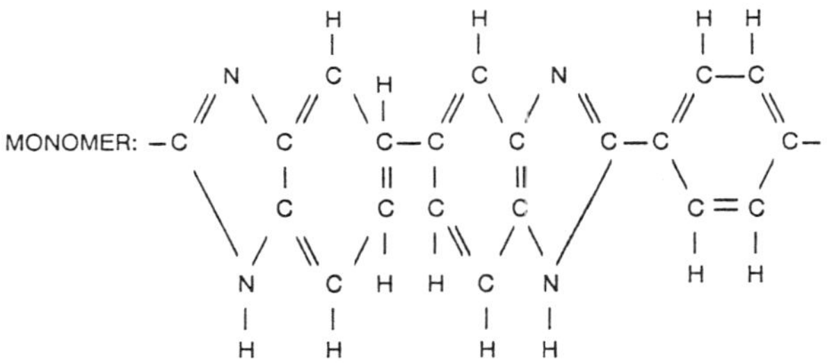

STRUCTURE: ring structure

APPLICATIONS: highest thermal stability of organic-based polymer makes it useful in fiber form

POLYBUTADIENE (BR)

GLASS TRANSITION TEMPERATURE: −102° to −10°C (−152° to 14°F) depending on amount of isotactic and syndiotactic present

HEAT-DISTORTION TEMPERATURE: 60°C (140°F)

MAXIMUM CONTINUOUS SERVICE TEMPERATURE: 70°C (158°F)

MINIMUM CONTINUOUS SERVICE TEMPERATURE: −70°C (−94°F)

MELTING/DECOMPOSITION TEMPERATURE: 90°–154°C (194°–309°F) depending on structure

SPECIFIC GRAVITY: 0.907

SPECIFIC HEAT: 1.63 kJ/kgK

COEFFICIENT OF LINEAR EXPANSION: 0.0008°K (0.00045°F)

ELONGATION: 500%

HARDNESS: 35–90 IRHD Shore A

TENSILE STRENGTH: 180 MPa (26,100 psi)

IMPACT STRENGTH: no break for notched or unnotched

FLEXURAL STRENGTH: 47 MPa (6,800 psi) good to low

FLEXURAL MODULUS: 560 MPa (81,200 psi)

TEAR STRENGTH: 140 MPa (20,300 psi)

ABRASION RESISTANCE: good to low temperature

ELECTRICAL RESISTIVITY: 10E16 Ohm/cm for volume, 10E16 Ohm for surface

DIELECTRIC STRENGTH: 94–106 kV/mm

DIELECTRIC CONSTANT: 2.7–2.9 at 1000 Hz

DISSIPATION FACTOR: 0.0031 at 1000 Hz

TRANSMISSION OF LIGHT: 91%

MOISTURE-VAPOR TRANSMISSION: 1.27X10E/s

GAS TRANSMISSION RATE: 3.57X10E-13 cm^3s/g for carbon dioxide, 8.8X10E-14 cm^3s/g for oxygen, 3.7X10E-12 cm^3s/g for ethylene oxide

CHEMICAL STABILITY; RESISTANCE TO:

ACIDS: poor

BASES: poor

SUNLIGHT: excellent

TOXICITY: suitable for food use

WEAKNESSES: poor tear resistance

MONOMER: 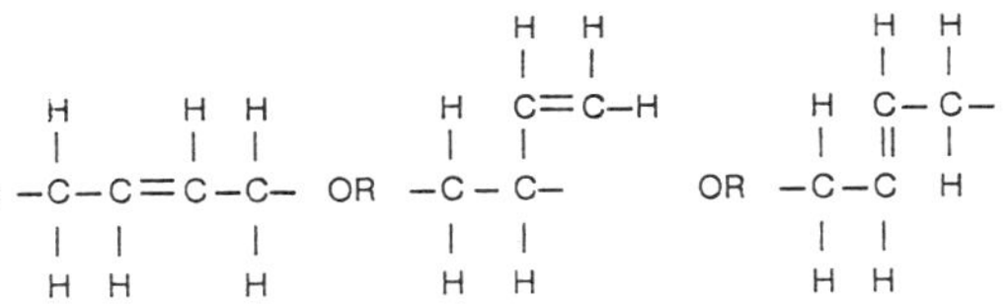

STRUCTURE: amorphous unless stretched; upon stretching, crystals will form and will dissolve again once the force is removed

APPLICATIONS: It is used in combination with SBR and NR for tires, since it retains its properties to low temperatures. It can also be used to make transparent bottles and films for food packaging. Given its tear strength, it is suitable for bubble packs.

POLYBUTYLENETEREPHTHALATE (PBT)

GLASS TRANSITION TEMPERATURE: 40°C (104°F)

HEAT-DISTORTION TEMPERATURE: 178°C (352°F)

MAXIMUM CONTINUOUS SERVICE TEMPERATURE: 100°C (212°F)

MINIMUM CONTINUOUS SERVICE TEMPERATURE: −30°C (−22°F)

MELTING/DECOMPOSITION TEMPERATURE: 224°C (435°F), melting

SPECIFIC GRAVITY: 1.31

SPECIFIC HEAT: 1.3 kJ/kgK

THERMAL CONDUCTIVITY: 0.21 W/mK

COEFFICIENT OF LINEAR EXPANSION: 0.00006°K (0.00033°F)

HARDNESS: 180 ball indention N/mm^2

ELONGATION: 15–150%

MODULUS OF ELASTICITY: 2000 MPa (290,000 psi)

TENSILE STRENGTH: 40 MPa (5800 psi)

TENSILE MODULUS: 2000–2300 MPa (290,000–330,000 psi)

IMPACT STRENGTH: no break for unnotched; 4 J/mm for notched

FLEXURAL STRENGTH: 96 MPa (14,000 psi)

FLEXURAL MODULUS: 2300–2400 MPa (330,000–350,000 psi)

CREEP: low; good fatigue resistance

COEFFICIENT OF FRICTION: 0.25

ELECTRICAL RESISTIVITY: 10E14 Ohm/m for volume; 10E13 Ohm for surface

DIELECTRIC STRENGTH: 42 kV/mm

DIELECTRIC CONSTANT: 3 at 50 Hz and 10E6 Hz

DISSIPATION FACTOR: 0.002 at 50 Hz; 0.02 at 10E6 Hz

TRANSMISSION OF LIGHT: low

WATER ABSORPTION: 0.08% of volume

FLAMMABILITY: burns

COMBUSTION PRODUCTS: carbon dioxide, carbon monoxide

CHEMICAL STABILITY; RESISTANCE TO:

ACIDS: moderate

BASES: excellent

SOLVENTS: good

OILS AND FATS: excellent

TOXICITY: suitable for food and cosmetic industry

MONOMER: 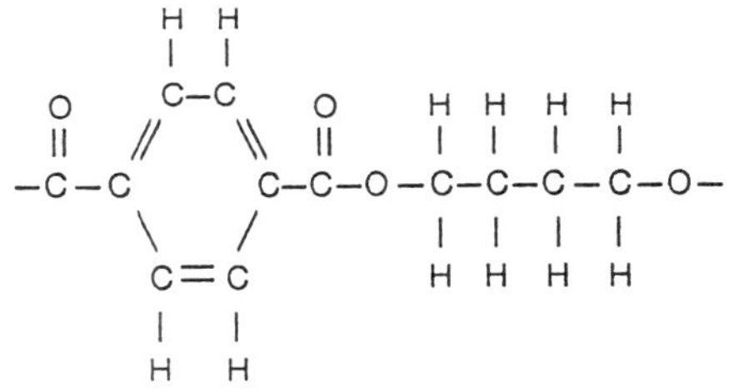

STRUCTURE: a member of the polyester family, very similar to PET, crystallizes to a high degree, thermoplastic

APPLICATIONS: automotive exterior parts, electrical applications

POLYCARBONATE (PC)

GLASS TRANSITION TEMPERATURE: 149°C (300°F)

HEAT-DISTORTION TEMPERATURE: 138°C (280°F)

MAXIMUM CONTINUOUS SERVICE TEMPERATURE: 135°C (275°F)

MINIMUM CONTINUOUS SERVICE TEMPERATURE: −100°C (−148°F)

MELTING/DECOMPOSITION TEMPERATURE:

140–150°C (284–302°F) for melting; 420–620°C (788–1148°F) for decomposition

SPECIFIC GRAVITY: 1.2

SPECIFIC HEAT: 1.17 kJ/kgK

THERMAL CONDUCTIVITY: 0.21 W/mK

COEFFICIENT OF LINEAR EXPANSION: 0.00006–0.00007°K (0.000035°F)

HARDNESS: 110 ball indention N/mm

ELONGATION: 100–130%

TENSILE MODULUS: 2,100–2,400 MPa; 300,000 psi

TENSILE STRENGTH: 56–67 MPa; 9000 psi

IMPACT STRENGTH: no break for unnotched; 20–30 J/mm for notched

FLEXURAL STRENGTH: 83 MPa (12,000 psi)

FLEXURAL MODULUS: 2,300–3,200 MPa; 320,000 psi

DEFLECTION TEMPERATURE UNDER FLEXURAL LOAD: 127°C (260°F) for a load of 1.8 MPa (264 psi); 138°C (280°F) for a load of 0.5 MPa (66 psi)

COMPRESSIVE STRENGTH: 83 MPa; 12,000 psi

ABRASION RESISTANCE: good

CREEP: under ambient conditions and an applied stress of 21 MPa (3,000 psi) the creep modulus is 2,380 MPa (345,000 psi) after 1 hr, 2,210 MPa (320,000 psi) after 100 hr and 2,140 MPa (310,000 psi) after 1000 hr; good high temperature creep properties

COEFFICIENT OF FRICTION: 0.38

ELECTRICAL RESISTIVITY: 10E17 Ohm/cm for volume; 10E15 Ohm for surface

DIELECTRIC STRENGTH: 38 kV/mm

DIELECTRIC CONSTANT: 3.0 at 50 Hz; 2.9 at 10E6 Hz

ARC RESISTANCE: 60 sec

DISSIPATION FACTOR: 0.0007 at 50 Hz; 0.01 at 10E6 Hz

REFRACTIVE INDEX: 1.58

TRANSMISSION OF LIGHT: 88%; often a blue tint; 2% haze

WATER ABSORPTION: 0.16% of volume

FLAMMABILITY: burns with orange flame

COMBUSTION PRODUCTS: black soot, carbon dioxide, carbon monoxide, phenol

FLASH IGNITION TEMPERATURE: 482°C (900°F)

BURNING RATE: self-extinguishing

CHEMICAL STABILITY; RESISTANCE TO
ACIDS: good
BASES: very poor
SOLVENTS: poor, can cause stress cracking
OILS AND FATS: excellent
SUNLIGHT: good
BACTERIAL AND FUNGAL ATTACK: will not support growth

TOXICITY: grades for food use available

WEAKNESSES: will not sustain continuous boiling in water

MONOMER:

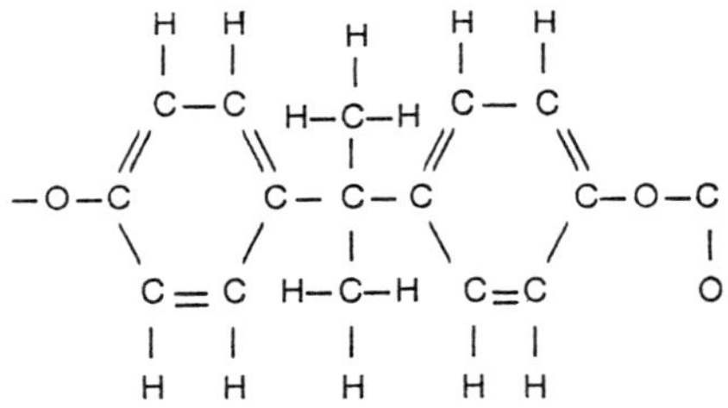

STRUCTURE: a thermoplastic which does not crystallize

APPLICATIONS: safety helmets, baby bottles due to high toughness, glazing

CELLULOSE ACETATE (CA)

GLASS TRANSITION TEMPERATURE: 157°C (315°F)

HEAT-DISTORTION TEMPERATURE: 90°C (194°F)

MAXIMUM CONTINUOUS SERVICE TEMPERATURE: 80°C (176°F)

MINIMUM CONTINUOUS SERVICE TEMPERATURE: −40°C (−40°F)

MELTING/DECOMPOSITION TEMPERATURE: 306°C (583°F) melting

SPECIFIC GRAVITY: 1.3

SPECIFIC HEAT: 1.6 KJ/kgK

THERMAL CONDUCTIVITY: 0.22 W/mK

COEFFICIENT OF LINEAR EXPANSION: .00012°K (0.000067°F); 0.000045°K (0.000025°F) for a foam of density 96 kg/m^3 (6 lb/ft^3)

HARDNESS: 50 ball-indention N/mm^2

ELONGATION: 4–55%; 25–28% for fiber form

TENSILE STRENGTH: 21–76 MPa (3,000–11,000 psi); 140–166 MPa (20,000–24,000 psi) for fiber form; 1.2 MPa (170 psi) for foam of density 96 kg/m^3 (6 lb/ft^3)

IMPACT STRENGTH: 65 J/mm; 5.3 J/mm (0.1 ft-lb) for unnotched foam of density 96 kg/m^3 (6 lb/ft^3)

TENSILE MODULUS: 690–2,760 MPa (100,000–400,000 psi)

FLEXURAL STRENGTH: 1 MPa (147 psi) for a foam of density 96 kg/m^3 (6 lb/ft^3)

FLEXURAL MODULUS: 38 MPa (5,500 psi) for a foam of density 96 kg/m^3 (6 lb/ft^3)

DEFLECTION TEMPERATURE UNDER FLEXURAL LOAD: 46–63°C (115–145°F) for a load of 1.8 MPa (264 psi); 49–98° C (120–209°F) under a load of 0.5 MPa (66 psi)

SHEAR STRENGTH: 1 MPa (140 psi) tension and 0.86 MPa (125 psi) compression for a foam of density 96 kg/m³ (6 lb/ft³)

COMPRESSIVE STRENGTH: 0.86 MPa (125 psi) for 96 kg/m³ (6 lb/ft³) density foam

ABRASION RESISTANCE: very good

COEFFICIENT OF FRICTION: 0.55

ELECTRICAL RESISTIVITY: 10E13 Ohm-cm by volume; 10E8 Ohm for surface

DIELECTRIC STRENGTH: 400 kV/cm

DIELECTRIC CONSTANT: 5.8 at 50 Hz, 4.6 at 10E6 Hz; 1.12 for 10E3–10E9 Hz and a foam density of 96 kg/m³ (6 lb/ft³)

ARC RESISTANCE: 100 sec

DISSIPATION FACTOR: 0.02 at 50 Hz, 0.03 at 10E6 Hz; 0.002–0.003 for 10E3–10E9 Hz and a foam density of 96 kg/m³ (6 lb/ft³)

REFLECTIVE INDEX: 1.50

TRANSMISSION OF LIGHT: high, optically clear

WATER ABSORPTION: 6% of volume, similar for foams

GAS TRANSMISSION RATE: 110 cc of oxygen through 1 mil film/24 hrs/100 sq in at 1 atm; 560 cc of carbon dioxide under the same conditions

FLAMMABILITY: burns readily with yellow flame

COMBUSTION PRODUCTS: carbon dioxide, carbon monoxide, acetic acid

FLASH IGNITION TEMPERATURE: 327°C (621°F)

LIMITING OXYGEN FOR SUSTAINED IGNITION: 25%

BURNING RATE: slow, 0.2–0.8 mm/s; 2mm/s (4.9 in/min) 96 kg/m³ (6 lb/ft³) foam

CHEMICAL STABILITY; RESISTANCE TO:
- *ACIDS:* very poor
- *BASES:* poor
- *SOLVENTS:* variable, excellent resistance to aliphatic hydrocarbons, ethers and petroleum
- *OILS AND FATS:* excellent
- *SUNLIGHT:* very good

RESPONSE TO IRRADIATION: degrades

WEAKNESSES: susceptible to static build-up

MONOMER: 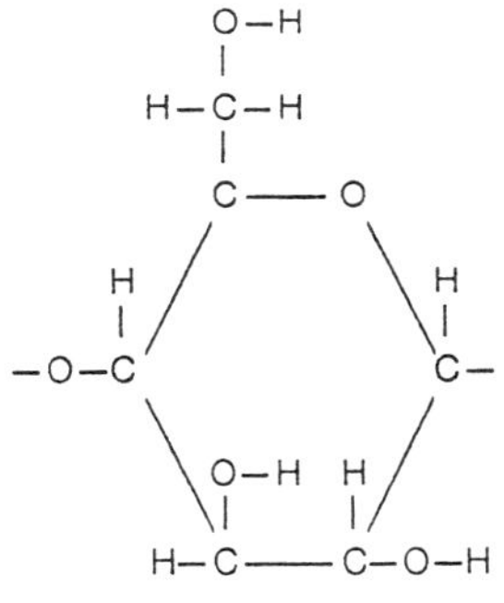

STRUCTURE: thermoplastic

APPLICATIONS: tool handles, electrical appliance housings, blister packs, pressure-sensitive tape, skylights, tape recorder ribbons

POLYCHLOROPRENE, neoprene (CR)

GLASS TRANSITION TEMPERATURE: −50°C (223°F)

MAXIMUM CONTINUOUS SERVICE TEMPERATURE: 88°C (190°F)

MINIMUM CONTINUOUS SERVICE TEMPERATURE: −20°C (−4°F)

MELTING/DECOMPOSITION TEMPERATURE: 70°C (158°F)

SPECIFIC GRAVITY: 1.23

ELONGATION: 400–800%

HARDNESS: 40–95 IRHD Shore A

TENSILE STRENGTH: 21 MPa (3000 psi)

ABRASION RESISTANCE: excellent

CHEMICAL STABILITY; RESISTANCE TO:
- *ACIDS:* good
- *BASES:* good
- *SOLVENTS:* variable; absorption of kerosene: 10–45%, gasoline: 10–45%, benzene: 100–400%, acetone: 15–50%, ethyl alcohol: 0–5%, carbon tetrachloride: 100–200%, lubricating oil: 0–150% volume increase in CR
- *OILS AND FATS:* good
- *SUNLIGHT:* very good
- *BACTERIAL AND FUNGAL ATTACK:* will not support growth

WEAKNESSES: difficult to process

MONOMER: 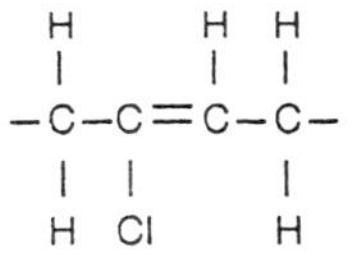

STRUCTURE: cross-linked rubber that does not crystallize

APPLICATIONS: machine mountings which are exposed to oil; in foam form it is used in divers' wet suits; durability encourages uses in conveyor belts and other tough environments; good damping qualities are helpful here as well

LOW DENSITY POLYETHYLENE, polythene (LDPE)

GLASS TRANSITION TEMPERATURE: −90°C (−130°F)

HEAT-DISTORTION TEMPERATURE: 40°C (104°F)

MAXIMUM CONTINUOUS SERVICE TEMPERATURE: 60–75°C (140–167°F)

MINIMUM CONTINUOUS SERVICE TEMPERATURE: −50°C (−58°F)

MELTING/DECOMPOSITION TEMPERATURE: 115°C (239°F) for melting; 340–440°C (644–824°F) for decomposition

SPECIFIC GRAVITY: 0.914–0.928

SPECIFIC HEAT: 2.1–2.5 kJ/kgK

THERMAL CONDUCTIVITY: 0.30 W/mK

COEFFICIENT OF LINEAR EXPANSION: 0.00025°K (0.00014°F)

HARDNESS: 13–20 ball indention N/mm^2

ELONGATION: 300–1000%; 25–50% for fiber form

TENSILE MODULUS: 140–3500 MPa; 20,000–200,000 psi

TENSILE STRENGTH: 80–230 MPa (12,000–34,000 psi); 80–186 MPa (12,000–27,000 psi) for fiber form; 0.14 MPa (20 psi) for 32 kg/m^3 (2 lb/ft^3) extruded foam and 0.69 MPa (100 psi) for compression molded foam

IMPACT STRENGTH: no break for both notched and unnotched

FLEXURAL STRENGTH: 5–21 MPa (725–3050 psi)

FLEXURAL MODULUS: 55–414 MPa (8000–60,000 psi)

DEFLECTION TEMPERATURE UNDER FLEXURAL LOAD: 32–41°C (90–105°F) for a load of 1.8 MPa (264 psi); 38–49°C (100–121°F) for a load of 0.5 MPa (66 psi)

COMPRESSIVE STRENGTH: 16 MPa (2300 psi); 0.06 MPa (8.5 psi) at 25% compression for a foam of 37 kg/m^3 2.3 lb/ft^3 density

ABRASION RESISTANCE: poor

COEFFICIENT OF FRICTION: 0.5

ELECTRICAL RESISTIVITY: 10E17 Ohm/cm for volume; 10E14 Ohm for surface

DIELECTRIC STRENGTH: 18–27 kV/mm

DIELECTRIC CONSTANT: 2.29 at 50 Hz, 2.28 at 10E6 Hz; 1.05 at 10E9 Hz and foam density of 32 kg/m^3 (2 lb/ft^3) 1.49 at 10E3 to 10E9 Hz and foam density of 512 kg/m^3 (32 lb/ft^3)

DISSIPATION FACTOR: 0.00015 at 50 Hz, 0.00008 at 10E6 Hz.; 0.0002 at 10E9 Hz and foam density of 32 kg/m^3 (2 lb/ft^3) 0.00033 at 10E3 to 10E9 Hz and foam density of 512 kg/m^3 (32 lb/ft^3)

REFRACTIVE INDEX: 1.51

TRANSMISSION OF LIGHT: high, milky white in color

MOISTURE-VAPOR TRANSMISSION: 21 g/mil/100 in^2-24 hr

WATER ABSORPTION: 0.01% of volume; up to 0.1% for foam form

GAS TRANSMISSION RATE: 573 cc of oxygen through 1 mil/24 hrs/100 in^2/atm; 1742 cc of carbon dioxide under the same conditions

FLAMMABILITY: will burn with blue flame

COMBUSTION PRODUCTS: no smoke, carbon dioxide, carbon monoxide

FLASH IGNITION TEMPERATURE: 343°C (650°F)

LIMITING OXYGEN FOR SUSTAINED IGNITION: 17%

BURNING RATE: very slow, 0.5 mm/s (1.04 in/min)

CHEMICAL STABILITY; RESISTANCE TO

ACIDS: excellent

BASES: excellent

SOLVENTS: poor to moderate

OILS AND FATS: good

SUNLIGHT: poor

BACTERIAL AND FUNGAL ATTACK: can support surface growth

TOXICITY: most grades meet FDA requirements for food use

RESPONSE TO IRRADIATION: at doses up to 30 megarads it cross-links

WEAKNESSES: susceptible to stress cracking and static build-up

MONOMER: 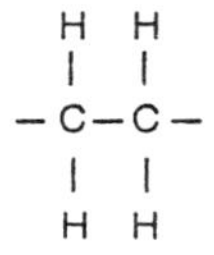

STRUCTURE: linear chain with occasional branching, crystallizable up to 65%, thermoplastic

APPLICATIONS: low temperature toughness and flexibility, as well as excellent chemical resistance, make it a good choice for chemical plant pipes. It is also used as packaging film.

HIGH DENSITY POLYETHYLENE, polyethylene (HDPE)

GLASS TRANSITION TEMPERATURE: −110°C (−166°F)

HEAT-DISTORTION TEMPERATURE: 60°C (140°F)

MAXIMUM CONTINUOUS SERVICE TEMPERATURE: 70–80°C (158–176°F)

MINIMUM CONTINUOUS SERVICE TEMPERATURE: −50°C (−58°F)

MELTING/DECOMPOSITION TEMPERATURE: 135°C (275°F) for melting; 340–440°C (644–824°F) for decomposition

SPECIFIC GRAVITY: 0.94–0.96

HEAT CAPACITY: 2.1–2.7 KJ/kgK

THERMAL CONDUCTIVITY: 0.38–0.51 W/mK

COEFFICIENT OF LINEAR EXPANSION: 0.0002°K (0.00011°F)

HARDNESS: 40–65 N/mm^2

ELONGATION: 15–100%; 10–20% for fiber form

TENSILE MODULUS: 700–1400 MPa (102,000–203,000 psi)

TENSILE STRENGTH: 21–38 MPa (3,100–5,500 psi); 340–590 MPa (49,000–86,000 psi) for fiber form; 0.14 MPa (20 psi) for 35 kg/m^3 (2.2 lb/ft^3) foam, 0.7 MPa (100 psi) for 144 kg/m^3 (9 lb/ft^3) foam, 9 MPa (1,300 psi) for 600 kg/m^3 (37 lbs/ft^3) density foam

EFFECT OF PRESSURE ON TENSILE PROPERTIES: under an applied pressure of 276 MPa (40,000 psi) the strain at rupture is 1.3 and the stress reaches a maximum of 54 MPa (7,800 psi) at 0.1 strain and then goes to zero at rupture

IMPACT STRENGTH: no break for unnotched and notched

FLEXURAL STRENGTH: 36 MPa (5,200 psi), 19 MPa (2,700 psi) for 600 kg/m^3 (37 lb/ft^3) density foam

FLEXURAL MODULUS: 1103 MPa (160,000 psi), 827 MPa (120,000 psi) for 600 kg/m^3 (37 lb/ft^3) density foam

DEFLECTION TEMPERATURE UNDER FLEXURAL LOAD: 43–54°C (110–130°F) for a load of 1.8 MPa (264 psi); 60–88°C (140–190°F) for a load of 0.5 MPa (66 psi)

SHEAR STRENGTH: 21 MPa (3,000 psi)

COMPRESSIVE STRENGTH: 17 MPa (2,400 psi); 0.06 MPa (8.5 psi) at 25% compression and 35 kg/m^3 (2.2 lb/ft^3) foam, 0.2 MPa (31 psi) for 25% compression and 144 kg/m^3 (9 lb/ft^3) foam

ABRASION RESISTANCE: poor

CREEP: under ambient conditions and an applied stress of 7 MPa (1,000 psi) the creep modulus is 345 MPa (50,000 psi) at 1 hr, 207 MPa (30,000 psi) at 100 hr and 172 MPa (25,000 psi) at 1000 hr

COEFFICIENT OF FRICTION: 0.25 against itself and austenitic steel

ELECTRICAL RESISTIVITY: 10E17 Ohm/cm for volume; 10E14 Ohm for surface

DIELECTRIC STRENGTH: 70 kV/mm

DIELECTRIC CONSTANT: 2.35 at 50 Hz; 2.34 at 10E6 Hz

DISSIPATION FACTOR: .00024 at 50 Hz; .00020 at 10E6 Hz

REFRACTIVE INDEX: 1.53

TRANSMISSION OF LIGHT: 30%; 40% haze, milky white in color

MOISTURE-VAPOR TRANSMISSION: 21 g/mil/100 in^2-24 hr

WATER ABSORPTION: 0.01% of volume; 0.5% of volume for foam

GAS TRANSMISSION RATE: 9X10E19 cm^{3}s/g for oxygen, 2X10E20 cm^{3}s/g for carbon dioxide

FLAMMABILITY: will burn with blue flame

COMBUSTION PRODUCTS: no smoke, carbon monoxide, carbon dioxide

FLASH IGNITION TEMPERATURE: 343°C (650°F)

LIMITING OXYGEN FOR SUSTAINED IGNITION: 17%

BURNING RATE: very slow, 0.4 mm/s (1–1.04 in/min)

CHEMICAL STABILITY; RESISTANCE TO:

ACIDS: excellent
BASES: excellent
SOLVENTS: good to excellent
OILS AND FATS: excellent
SUNLIGHT: poor, crazes
BACTERIAL AND FUNGAL ATTACK: will not support growth

TOXICITY: most grades approved for food use

RESPONSE TO IRRADIATION: at doses up to 30 megarads HDPE cross-links

WEAKNESSES: susceptible to stress cracking and static build-up

MONOMER: 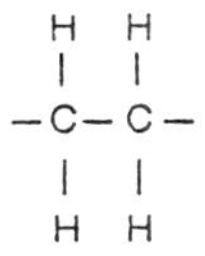

STRUCTURE: thermoplastic, linear chain which is crystallizable to 90%

APPLICATIONS: crates and pallets, bottles, toys and housewares

POLYETHYLENETEREPHTHALATE, dacron, mylar (PET)

GLASS TRANSITION TEMPERATURE: 67°C (153°F)

HEAT-DISTORTION TEMPERATURE: 180°C (356°F)

MAXIMUM CONTINUOUS SERVICE TEMPERATURE: 100°C (212°F)

MINIMUM CONTINUOUS SERVICE TEMPERATURE: −20°C (−4°F)

MELTING/DECOMPOSITION TEMPERATURE: 265°C (509°F) highly crystalline polymer melts at 271°C (520°F)

SPECIFIC GRAVITY: 1.37

SPECIFIC HEAT: 1.05 kJ/kgK

THERMAL CONDUCTIVITY: 0.24 W/mK

COEFFICIENT OF LINEAR EXPANSION: 0.00007°K (0.00004°F)

HARDNESS: 200 ball indention N/mm^2

ELONGATION: 50–300%

TENSILE MODULUS: 3100 MPa (450,000 psi)

TENSILE STRENGTH: 47 MPa (6,820 psi)

IMPACT STRENGTH: no break for unnotched; 4 J/mm for notched

FLEXURAL STRENGTH: 112 MPa (16,300 psi)

ABRASION RESISTANCE: good, 3 mg/1000c

COEFFICIENT OF FRICTION: 0.25

ELECTRICAL RESISTIVITY: 10E16 at 50 Hz; 10E16 at 10E6 Hz

DIELECTRIC STRENGTH: 42 kV/mm

DISSIPATION FACTOR: 0.002 at 50 Hz; 0.02 at 10E6 Hz

MOISTURE-VAPOR PERMEABILITY: 48 g/mil/100 in^2-24 hr

WATER ABSORPTION: 0.3% of volume

FLAMMABILITY: ignites

COMBUSTION PRODUCTS: carbon dioxide, carbon monoxide

BURNING RATE: self-extinguishing

CHEMICAL STABILITY; RESISTANCE TO:
ACIDS: very good
BASES: good
SOLVENTS: good to excellent
OILS AND FATS: excellent

WEAKNESSES: difficult to mold

MONOMER:

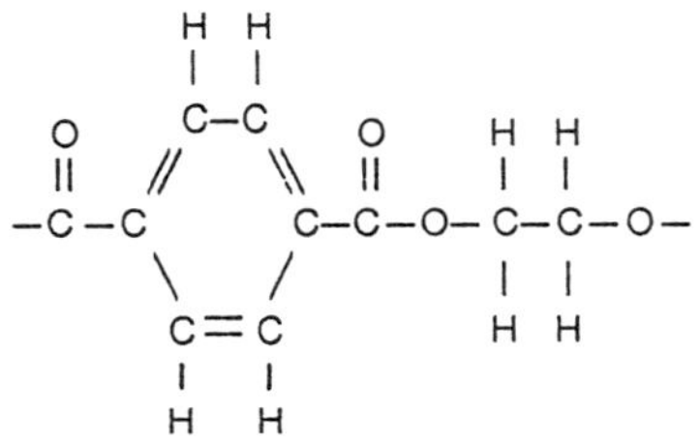

STRUCTURE: a family of thermoplastic polyesters that can crystallize up to 65%

APPLICATIONS: Biaxially oriented film is used in magnetic tape. Bottles for carbonated beverages also constitute part of its market. Most of it is used in fiber form for apparel.

POLY(HEXAMETHYLENE ADIPAMIDE), polyamide-6,6 or nylon-6,6 (PA)

GLASS TRANSITION TEMPERATURE: 47°C (117°F)

HEAT-DISTORTION TEMPERATURE: 200°C (392°F)

MAXIMUM CONTINUOUS SERVICE TEMPERATURE: 80–120°C (176–248°F)

MINIMUM CONTINUOUS SERVICE TEMPERATURE: −30°C (−22°F)

MELTING/DECOMPOSITION TEMPERATURE: 265°C (509°F) for melting; 300–400°C (572–752°F) for decomposition

SPECIFIC GRAVITY: 1.14

SPECIFIC HEAT: 1.7 kJ/kgK

THERMAL CONDUCTIVITY: 0.23 W/mK

COEFFICIENT OF LINEAR EXPANSION: 0.00008°K (0.000045°F)

HARDNESS: 100 ball indention N/mm^2

ELONGATION: 150–300%; 19–32% for fiber form

TENSILE MODULUS: 2,000 MPa; 200,000–400,000 psi

TENSILE STRENGTH: 77–84 MPa (11,000–12,000 psi); 450–850 MPa (65,000–123,000 psi) for fiber form

COMPRESSIVE STRENGTH: 90 MPa (13,000 psi)

IMPACT STRENGTH: no break for unnotched; 15–20 J/mm for notched

FLEXURAL STRENGTH: 95 MPa (13,800 psi)

FLEXURAL MODULUS: 1,200–3,100 MPa (174,000–450,000 psi)

DEFLECTION TEMPERATURE UNDER FLEXURAL LOAD: 48–90°C (118–194°F) for a load of 1.8 MPa (264 psi); 68–235°C (154–455°F) for a load of 0.5 MPa (66 psi)

CREEP: under ambient conditions and an applied stress of 10.3 MPa (1,500 psi) the creep modulus is 1,103 MPa (160,000 psi) after 1 hr, 713 MPa (115,000 psi) after 100 hr and 690 MPa (100,000 psi) after 1000 hr

COEFFICIENT OF FRICTION: 0.3, 0.35 against austenitic steel

ELECTRICAL RESISTIVITY: 10E12 Ohm/cm for volume; 10E10 Ohm for surface

DIELECTRIC STRENGTH: 60 kV/mm

DIELECTRIC CONSTANT: 8.0 at 50 Hz; 4.0 at 10E6 Hz

DISSIPATION FACTOR: 0.14 at 50 Hz; 0.08 at 10E6 Hz

REFRACTIVE INDEX: 1.53

TRANSMISSION OF LIGHT: variable, usually off-white and opaque

WATER ABSORPTION: 2.5% of volume

FLAMMABILITY: ignites

COMBUSTION PRODUCTS: soot, carbon dioxide, carbon monoxide, NH3, amines

FLASH IGNITION TEMPERATURE: 424°C (795°F)

LIMITING OXYGEN FOR SUSTAINED IGNITION: 29%

BURNING RATE: slow to self-extinguishing

CHEMICAL STABILITY; RESISTANCE TO:
ACIDS: very poor

BASES: excellent
SOLVENTS: excellent
OILS AND FATS: excellent
SUNLIGHT: poor, discolors
BACTERIAL AND FUNGAL ATTACK: will not support growth

RESPONSE TO IRRADIATION: cross-links

WEAKNESSES: dimensional instability due to absorption of water, electrical properties vary with moisture content, subject to static build-up

MONOMER: 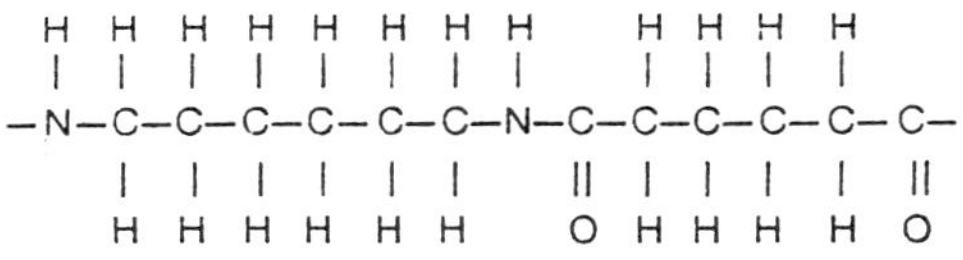

STRUCTURE: thermoplastic with a maximum crystallinity of 65%

APPLICATIONS: gears, bushes, cams, and bearings due to high toughness even at low temperature. In fiber form it is used extensively in home furnishings, apparel and tire cord.

POLYISOBUTYLENE (PIB)

GLASS TRANSITION TEMPERATURE: −70°C (−94°F)

MAXIMUM CONTINUOUS SERVICE TEMPERATURE: 65°C (149°F)

MINIMUM CONTINUOUS SERVICE TEMPERATURE: −50°C (−58°F)

MELTING/DECOMPOSITION TEMPERATURE: 288–425°C (550–880°F) for decomposition

SPECIFIC GRAVITY: 0.91–0.93

SPECIFIC HEAT: 1.68 kJ/kgK

THERMAL CONDUCTIVITY: 0.12–0.2 W/mK

COEFFICIENT OF LINEAR EXPANSION: 0.00012°K (0.000067°F)

ELONGATION: 1000%

TENSILE STRENGTH: 2–6 MPa

IMPACT STRENGTH: no break for both notched and unnotched tests

ELECTRICAL RESISTIVITY: 10E15 Ohm/cm for volume; 10E13 for surface

DIELECTRIC CONSTANT: 2.3 at 50 Hz

DISSIPATION FACTOR: 0.0004 at 50 Hz

TRANSMISSION OF LIGHT: low

WATER ABSORPTION: 0.01% of volume

CHEMICAL STABILITY; RESISTANCE TO:
SUNLIGHT: excellent

RESPONSE TO IRRADIATION: degrades

MONOMER: 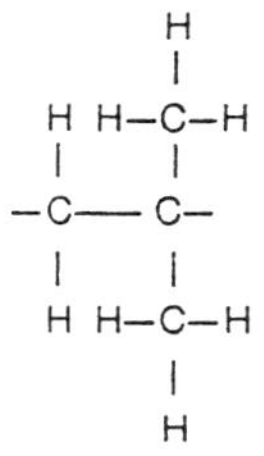

STRUCTURE: usually the methyl groups would inhibit rotation of the chain, but since all configurations are strained, the chain rotates freely and the result is an effective elastomer; it does not crystallize extensively

APPLICATIONS: elastic properties are used for specialized use

POLY(ISOBUTYLENE-ISOPRENE) COPOLYMER, butyl (BUTYL)

GLASS TRANSITION TEMPERATURE: −79°C (−110°F)

MAXIMUM CONTINUOUS SERVICE TEMPERATURE: 100°C (212°F)

MINIMUM CONTINUOUS SERVICE TEMPERATURE: −54°C (−65°F)

SPECIFIC GRAVITY: 0.92

HARDNESS: 40–75 Shore A

ELONGATION: 100–700%

TENSILE STRENGTH: 10.3 MPa (1500 psi)

ABRASION RESISTANCE: good

DIELECTRIC CONSTANT: 2.35 at 10E6 Hz

GAS PERMEABILITY: extremely low

CHEMICAL STABILITY; RESISTANCE TO:
ACIDS: excellent
BASES: good
SOLVENTS: variable; absorption of kerosene: 100–800%, gasoline: 150–400%, benzene: 100–200%, acetone: 0–10%, ethyl alcohol: 5–20%, carbon tetrachloride: 200–500%, lubricating oil: 500% increase in volume of butyl
OILS AND FATS: excellent
SUNLIGHT: very good

TOXICITY: medical application when carbon-black is omitted

MONOMER: 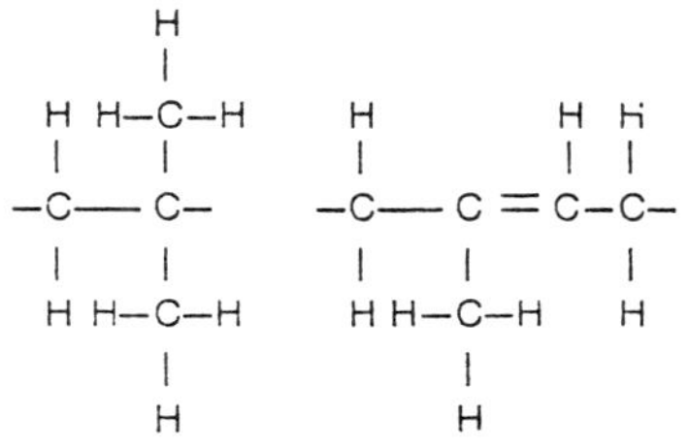

STRUCTURE: cross-linked elastomer which does not crystallize

APPLICATIONS: inner tubes and tire liners due to excellent air retention

cis-POLYISOPRENE, natural rubber (NR)

GLASS TRANSITION TEMPERATURE: −72°C (−98°F)

MAXIMUM CONTINUOUS SERVICE TEMPERATURE: 82°C (180°F)

MINIMUM CONTINUOUS SERVICE TEMPERATURE: −54°C (−65°F)

MELTING/DECOMPOSITION TEMPERATURE: 36°C (97°F) but on stretching crystals can raise the melting point to as high as 100°C (212°F)

SPECIFIC GRAVITY: 0.93

SPECIFIC HEAT: 1.59 kJ/kgK

COEFFICIENT OF LINEAR EXPANSION: 0.00062°K (0.00035°F)

HARDNESS: 30 ball indention (IRHD)

ELONGATION: 470% for vulcanized, 1200% for raw form

TENSILE MODULUS: 0.9 MPa (130 psi)

TENSILE STRENGTH: vulcanized rubber is greatly dependent on crystal formation for its strength and is thus very temperature sensitive, the strength is about 6.9 MPa (1,000 psi) at 0° and 40°C (32° and 104°F) with a bell-shaped function between these two temperatures which shows a maximum at about 20°C (68°F) and 22.8 MPa (3,300 psi); 2.1 MPa (300 psi) for raw form

SHEAR MODULUS: 0.3 MPa (43 psi)

ABRASION RESISTANCE: excellent

GAS TRANSMISSION RATE: fairly high

CHEMICAL STABILITY; RESISTANCE TO:
- *ACIDS:* fair
- *SOLVENTS:* poor, unvulcanized is soluble
- *OILS AND FATS:* poor
- *SUNLIGHT:* poor

TOXICITY: unvulcanized material used for artificial hearts

RESPONSE TO IRRADIATION: causes cross-linking

MONOMER:

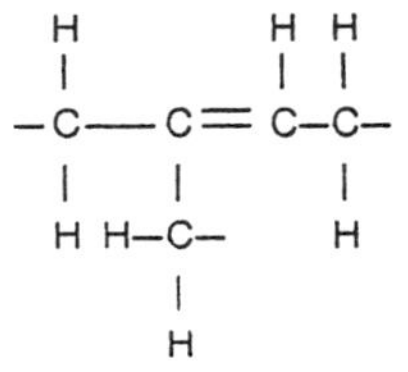

STRUCTURE: always amorphous unless stretched, during which crystallization can occur, the crystals will dissolve once the force is removed

APPLICATIONS: tires, antivibration mountings, dock fenders due to good tear resistance and compression set resistance. NR also adheres very well to fabrics for reinforcement.

POLY(MELAMINE-FORMALDEHYDE), melamine (MF)

GLASS TRANSITION TEMPERATURE: decomposes first

HEAT-DISTORTION TEMPERATURE: 200°C (392°F)

MAXIMUM CONTINUOUS SERVICE TEMPERATURE: 80°C (176°F)

SPECIFIC GRAVITY: 1.5

SPECIFIC HEAT: 1.2 kJ/kgK

THERMAL CONDUCTIVITY: 0.5 W/mK

COEFFICIENT OF LINEAR EXPANSION: 0.00005°K (0.00003°F)

HARDNESS: 400 ball indention N/mm^2

ELONGATION: 0.9%

TENSILE MODULUS: 9100 MPa (1,400,000 psi)

TENSILE STRENGTH: 30 MPa (4400 psi)

IMPACT STRENGTH: 7 J/mm (130 ft-lb/in) for unnotched; 1.5 J/mm (28 ft-lb/in) for notched

FLEXURAL STRENGTH: 117 MPa (17,000 psi)

FLEXURAL MODULUS: 8000–10,000 MPa (1,160,000–1,450,000 psi)

COMPRESSIVE STRENGTH: 207 MPa (30,000 psi)

ELECTRICAL RESISTIVITY: 10E11 Ohm/cm for volume; 10E8 Ohm for surface

DIELECTRIC STRENGTH: 30 kV/mm

DIELECTRIC CONSTANT: 9 at 50 Hz; 8 at 10E6 Hz

ARC RESISTANCE: 180 sec.

DISSIPATION FACTOR: 0.06 at 50 Hz; 0.03 at 10E6 Hz

TRANSMISSION OF LIGHT: low, gray and opaque

WATER ABSORPTION: 0.1–0.6% of volume

FLASH IGNITION TEMPERATURE: 602°C (1,116°F)

BURNING RATE: self-extinguishing

CHEMICAL STABILITY; RESISTANCE TO:
- *ACIDS:* very poor
- *BASES:* poor
- *SOLVENTS:* good to excellent
- *OILS AND FATS:* excellent
- *SUNLIGHT:* good

RESPONSE TO IRRADIATION: cross-links

MONOMER:

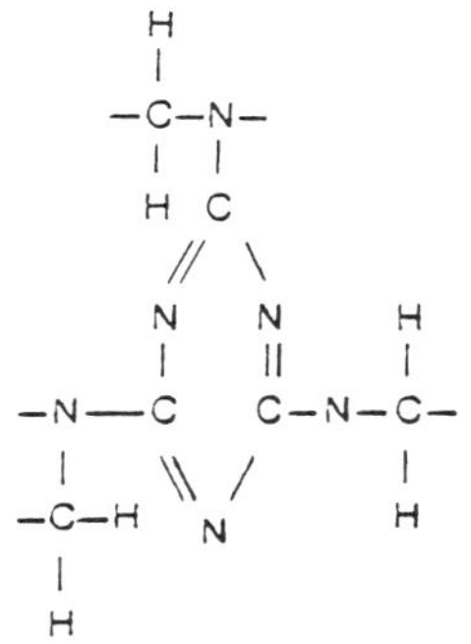

STRUCTURE: cross-linked thermoset that does not crystallize

APPLICATIONS: good stain resistance makes it useful for tableware and picnicware

POLY(METHYL METHACRYLATE) ACRYLIC (PMMA)

GLASS TRANSITION TEMPERATURE: 100°C (212°F)

HEAT-DISTORTION TEMPERATURE: 70–100°C (160–212°F)

MAXIMUM CONTINUOUS SERVICE TEMPERATURE: 65–90°C (150–195°F)

MINIMUM CONTINUOUS SERVICE TEMPERATURE: −40°C (−40°F)

MELTING/DECOMPOSITION TEMPERATURE: 160°C (320°F) melting

SPECIFIC GRAVITY: 1.17–1.2

SPECIFIC HEAT: 1.47 KJ/kgK

THERMAL CONDUCTIVITY: 0.18 W/mK

COEFFICIENT OF LINEAR EXPANSION: 0.00007°K (0.00004°F)

HARDNESS: 180–200 ball indention N/mm^2

ELONGATION: 2–10%

TENSILE MODULUS: 2700–3200 MPa (392,000–465,000 psi)

TENSILE STRENGTH: 50–77 MPa (7,300–11,000 psi); 1.6 MPa (235 psi) for a foam of density 50 kg/m^3 (3.12 lb/ft^3), 3 MPa (455 psi) for a foam of density 100 kg/m^3 (6.24 lb/ft^3)

EFFECT ON TENSILE PROPERTIES OF TEMPERATURE: at −196°C (−319°F) the strain at failure is 2% and the stress is 124 MPa (18,000 psi); at −60°C (−76°F) the strain at failure is 3% and the stress is 124 MPa (17,000 psi); at −20°C (−4°F) the strain at failure is 4% and the stress is 103 MPa (15,000 psi); at 40°C (104°F) the strain at failure is 13.5% and the stress goes to zero, the maximum stress of 41 MPa (6,000 psi) being at 5% strain; at 100°C (212°F) the stress is a maximum at 2% strain and holds nearly steady until failure when it goes to zero and the strain is 184%

IMPACT STRENGTH: 18 J/mm (337 ft-lb/in) for unnotched; 2 J/mm (37 ft-lb/in) for notched

FLEXURAL STRENGTH: 117 MPa (17,000 psi); 1.7 MPa (240 psi) for a foam of density 50 kg/m^3 (3.12 lb/ft^3) 3.9 MPa (560 psi) for a foam of density 100 kg/m^3 (6.24 lb/ft^3)

FLEXURAL MODULUS: 3,100 MPa (450,000 psi)

DEFLECTION TEMPERATURE UNDER A FLEXURAL LOAD: 85–93°C (185–200°F) under a load of 1.8 MPa (264 psi); 95–107°C (203–225°F) under a load of 0.5 MPa (66 psi)

COMPRESSIVE STRENGTH: 110 MPa (16,000 psi); 1 MPa (142 psi) for a foam of density 50 kg/m^3 (3.12 lb/ft^3), 2.8 MPa (406 psi) for a foam of density 100 kg/m^3 (6.24 lb/ft^3)

ABRASION RESISTANCE: poor

CREEP: under an applied stress of 10 MPa (1,500 psi) the creep modulus is 2,760 MPa (400,000 psi) at 1 hr, 2,240 MPa (325,000 psi) at 100 hr; for an applied stress of 30 MPa (4,350 psi) the creep modulus is 2,340 MPa (340,000 psi) at 1 hr and 1,620 MPa (235,000 psi) at 100 hr

FATIGUE: excellent; able to withstand cycling between a tensile and compressive load of 35 MPa (5,000 psi) indefinitely

COEFFICIENT OF FRICTION: 0.4–0.6, 0.54 against austenitic steel

ELECTRICAL RESISTIVITY: 10E15 Ohm/cm for volume; 10E15 Ohm for surface

DIELECTRIC STRENGTH: 40–50 KV/mm

DIELECTRIC CONSTANT: 3.3–3.9 at 50 Hz; 2.2–3.2 at 10E6 Hz

ARC RESISTANCE: no track

DISSIPATION FACTOR: 0.04–0.06 at 50 Hz; 0.004–0.04 at 10E6 Hz

REFRACTIVE INDEX: 1.49

TRANSMISSION OF LIGHT: 92%, colorless, 3% haze

SOUND ABSORPTION: 23 at 530 Hz, 20 at 620 Hz, 5 at 3200 Hz, 15 at 20,000 Hz, 15 at 60,000 Hz for a foam

MOISTURE-VAPOR PERMEABILITY: low

WATER ABSORPTION: 0.1–0.4% of volume

FLAMMABILITY: high; produces blue flame

COMBUSTION PRODUCTS: carbon dioxide, carbon monoxide

FLASH IGNITION TEMPERATURE: 338°C (640°F)

LIMITING OXYGEN FOR SUSTAINED IGNITION: 17%

BURNING RATE: slow, 0.2–0.8 mm/s (0.5–2 in/min)

CHEMICAL STABILITY; RESISTANCE TO:
- *ACIDS:* good
- *BASES:* excellent
- *SOLVENTS:* good except halogen alkanes and esters
- *OILS AND FATS:* excellent
- *SUNLIGHT:* excellent
- *BACTERIAL AND FUNGAL ATTACK:* will not support growth

TOXICITY: used as molded implants for skull

RESPONSE TO IRRADIATION: degrades

WEAKNESSES: stress cracking and static build-up

MONOMER:

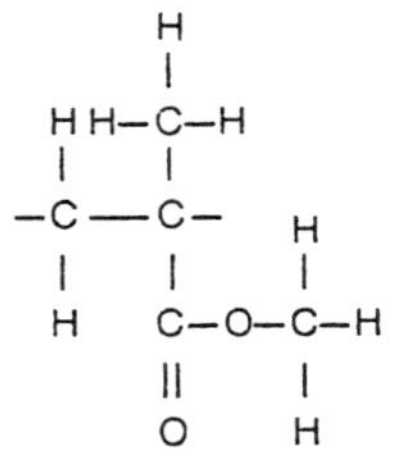

STRUCTURE: a thermoplastic which does not crystallize or branch

APPLICATIONS: optical quality lenses due to clarity; non-lubricated gears and bearings due to strength, rigidity, dimensional stability and good frictional properties; excellent moldability allows for a snap-fit

POLYOXYMETHYLENE, acetal (POM)

GLASS TRANSITION TEMPERATURE: −75°C (−103°F)

HEAT-DISTORTION TEMPERATURE: 160–173°C (320–343°F)

MAXIMUM CONTINUOUS SERVICE TEMPERATURE: 90–110°C (194–230°F)

MINIMUM CONTINUOUS SERVICE TEMPERATURE: −60°C (−76°F)

MELTING/DECOMPOSITION TEMPERATURE: 175°C (347°F)

SPECIFIC GRAVITY: 1.41–1.42

SPECIFIC HEAT: 1.46 kJ/kgK

THERMAL CONDUCTIVITY: 0.25–0.30 W/mK

COEFFICIENT OF LINEAR EXPANSION: 0.00009–0.00011°K (0.00005–0.00006°F)

HARDNESS: 150–170 ball indention N/mm²

ELONGATION: 25–70%

TENSILE MODULUS: 2800–3200 MPa (410,000–470,000 psi)

TENSILE STRENGTH: 62–70 MPa (9,000–10,000 psi)

IMPACT STRENGTH: 100 J/mm for unnotched; 8 J/mm for notched

FLEXURAL STRENGTH: 97 MPa (14,100 psi)

FLEXURAL MODULUS: 2,600–2,800 MPa (380,000–410,000 psi)

DEFLECTION TEMPERATURE UNDER FLEXURAL LOAD: 136°C (277°F) for a load of 1.8 MPa (264 psi); 172°C (342°F) for a load of 0.5 MPa (66 psi)

COMPRESSIVE STRENGTH: 124 MPa (18,000 psi)

ABRASION RESISTANCE: good

CREEP: for an applied stress of 10 MPa (1,500 psi) at ambient the creep modulus is 2,700 MPa (390,000 psi) in 1 hr, 1,930 MPa (280,000 psi) in 100 hr and 1,730 MPa (250,000 psi) in 1000 hr

FATIGUE: loaded to 35 MPa (5,000 psi), there was no failure after 10E6 cycles

COEFFICIENT OF FRICTION: 0.21–0.35

ELECTRICAL RESISTIVITY: 10E15 Ohm/cm for volume; 10E13 Ohm for surface

DIELECTRIC STRENGTH: 38–50 kV/mm

DIELECTRIC CONSTANT: 3.7 at 50 and 10E6 Hz

ARC RESISTANCE: 125 sec.

DISSIPATION FACTOR: 0.005 at 50 and 10E6 Hz

REFRACTIVE INDEX:1.48

TRANSMISSION OF LIGHT: low

WATER ABSORPTION: 0.22–0.25% of volume

FLAMMABILITY: moderate; blue flame

COMBUSTION PRODUCTS: no smoke, carbon dioxide, carbon monoxide

LIMITING OXYGEN FOR SUSTAINED IGNITION: 16%

BURNING RATE: slow, 0.42 mm/s (1 in/min)

CHEMICAL STABILITY; RESISTANCE TO:
- *ACIDS:* poor
- *BASES:* excellent
- *SOLVENTS:* excellent
- *OILS AND FATS:* excellent
- *SUNLIGHT:* poor

TOXICITY:approved for use with consumable water

WEAKNESSES: difficult to bond

MONOMER:

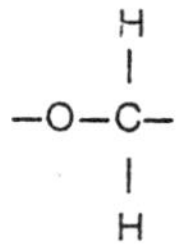

STRUCTURE: thermoplastic, maximum crystallinity is 85%

APPLICATIONS: pipe fittings, gears; bearings due to low coefficient of friction and good resistance to fatigue

POLY(PHENOL-FORMALDEHYDE); phenolic (PF)

GLASS TRANSITION TEMPERATURE: decomposes without transforming

HEAT-DISTORTION TEMPERATURE: 112–125°C (234–257°F)

MAXIMUM CONTINUOUS SERVICE TEMPERATURE: 110°C (230°F)

MINIMUM CONTINUOUS SERVICE TEMPERATURE: no practical limit for solid or foam

MELTING/DECOMPOSITION TEMPERATURE: decomposes at 220°C (428°F) for foam

SPECIFIC GRAVITY: 1.4

SPECIFIC HEAT: 1.3 KJ/kgK

THERMAL CONDUCTIVITY: 0.35 W/mK

COEFFICIENT OF LINEAR EXPANSION: 0.00003–0.00005°K (0.00002–0.00003°F)

HARDNESS: 250–320 ball hardness N/mm²

ELONGATION 0.4–0.8%

TENSILE MODULUS: 5600–12,000 MPa (810,000–1,700,000 psi)

TENSILE STRENGTH: 25 MPa (3,600 psi); 0.06–0.1 MPa (9–15 psi) for a foam of density 18 kg/m³ (1.1 lb/ft³), 0.1 MPa (15 psi) for a foam of density 42 kg/m³ (2.6 lb/ft³)

IMPACT STRENGTH: 6 J/mm for unnotched; 1.5 J/mm for notched

FLEXURAL STRENGTH: 83–103 MPa (12,000–15,000 psi)

FLEXURAL MODULUS: 14,000 MPa (2,000,000 psi) for glass-fiber filled

COMPRESSIVE STRENGTH: 69–207 MPa (10,000–30,000 psi); 0.007 MPa (10 psi) for a foam of 18 kg/m³ (1.1 lb/ft³) density, 0.2 MPa (29 psi) for a foam of 42 kg/m³ (2.6 lb/ft³) density

ABRASION RESISTANCE: very good

CREEP: under ambient conditions and an applied stress of 14 MPa (2,000 psi) the creep modulus is 30,000 MPa (4,400,000 psi) after 1 hr, 28,960 MPa (4,200,000 psi) after 100 hr and 47,580 MPa (4,000,000 psi) after 1000 hr for glass-fiber filled

COEFFICIENT OF FRICTION: 0.61

ELECTRICAL RESISTIVITY: 10E11 Ohm/cm for volume; 10E8 for surface

DIELECTRIC STRENGTH: 30–40 kV/mm

DIELECTRIC CONSTANT: 6 at 50 Hz; 4.5 at 10E6 Hz

ARC RESISTANCE: 100 sec

DISSIPATION FACTOR: 0.1 at 50 Hz; 0.03 at 10E6 Hz

REFRACTIVE INDEX: 1.5–1.7

TRANSMISSION OF LIGHT: low

SOUND ABSORPTION: Absorption coefficient: for a foam of density 37–64 kg/m³ (2.3–4 lb/ft³) and 2.5 cm (1 in) thickness, 0.05 at 125 Hz, 0.25 at 250 Hz, 0.65 at 500 Hz, 1.00 at 1000 Hz

WATER ABSORPTION: 0.3–1.2% of volume

FLAMMABILITY: flame spreading rate of 10

COMBUSTION PRODUCTS: very little smoke, carbon dioxide, carbon monoxide, HCOOH

FLASH IGNITION TEMPERATURE: 482°C (900°F)

BURNING RATE: self-extinguishing in both solid and foam form

CHEMICAL STABILITY; RESISTANCE TO:
- *ACIDS:* can decompose especially in foam form
- *BASES:* can decompose especially in foam form
- *SOLVENTS:* very good
- *OILS AND FATS:* excellent
- *SUNLIGHT:* darkens
- *BACTERIAL AND FUNGAL ATTACK:* will not support growth

WEAKNESSES: brittle

MONOMER:

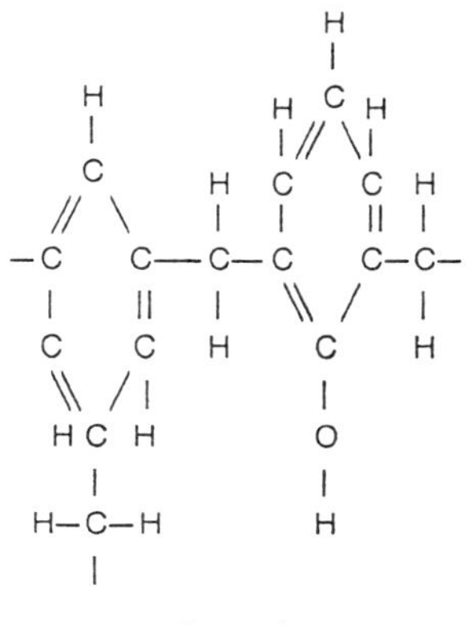

STRUCTURE: thermosetting and does not crystallize, cross-linked

APPLICATIONS: distributor heads; electrical plugs, switches and sockets; applications that require heat and dimensional stability; agricultural uses because in foam form it can retain liquids due to open-cell structure

POLY(PHENYLENE OXIDE) (PPO)

GLASS TRANSITION TEMPERATURE: 90–220°C (194–428°F)

HEAT DISTORTION TEMPERATURE: 148°C (298°F)

MAXIMUM CONTINUOUS SERVICE TEMPERATURE: 80°C (176°F)

MINIMUM CONTINUOUS SERVICE TEMPERATURE: −30°C (−22°F)

MELTING/DECOMPOSITION TEMPERATURE: 262–484°C (504–903°F) for melting

SPECIFIC GRAVITY: 1.06

SPECIFIC HEAT: 1.4 kJ/kgK

THERMAL CONDUCTIVITY: 0.23 W/mK

COEFFICIENT OF LINEAR EXPANSION: 0.00006°K (0.00003°F)

HARDNESS: R115 Rockwell

ELONGATION: 50–60%

TENSILE MODULUS: 2500 MPa (360,000 psi)

TENSILE STRENGTH: 55–68 MPa; 8000–10,000 psi

IMPACT STRENGTH: no break for unnotched; 214–534 J/m (4–10 ft-lb/in) notched

FLEXURAL STRENGTH: 93 MPa (13,500 psi)

FLEXURAL MODULUS: 1,800–3,300 MPa (360,000–480,000 psi)

COMPRESSIVE STRENGTH: 110 MPa (16,000 psi)

ABRASION RESISTANCE: very good

CREEP: under ambient conditions and an applied stress of 10 MPa (1,500 psi) the creep modulus is 2,970 MPa (430,000 psi) at 1 hr, 2,760 MPa (400,000 psi) at 100 hr, and 2,200 MPa (320,000 psi) at 1000 hr

ELECTRICAL RESISTIVITY: 10E16 Ohm/cm for volume; 10E14 Ohm for surface

DIELECTRIC STRENGTH: 17.7 kV/mm (450 V/mil)

DIELECTRIC CONSTANT: 2.6 at 50 Hz and 10E6 Hz

ARC RESISTANCE: 75 sec.

DISSIPATION FACTOR: 0.0004 at 50 Hz; 0.0009 at 10E6 Hz

TRANSMISSION OF LIGHT: low; tan in color

WATER ABSORPTION: 0.06% of volume

FLAMMABILITY: difficult to ignite

BURNING RATE: self-extinguishing

CHEMICAL STABILITY; RESISTANCE TO:
- *ACIDS:* excellent
- *BASES:* excellent
- *SOLVENTS:* very poor
- *SUNLIGHT:* colors fade

MONOMER:

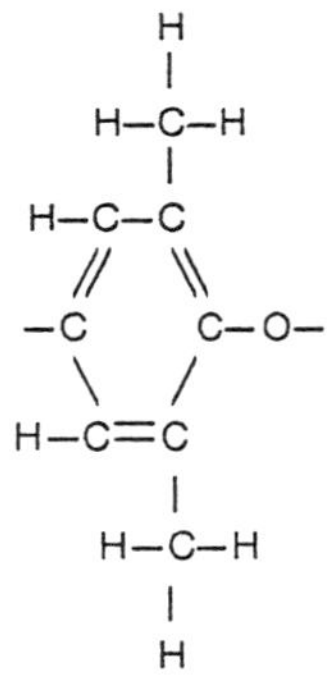

STRUCTURE: thermoplastic

APPLICATIONS: exterior automotive parts due to high impact strength even at low temperature, computer housings, television cases; pump impellers due to high fatigue endurance limit

POLY(PHENYLENE SULPHIDE) (PPS)

GLASS TRANSITION TEMPERATURE: 193–204°C (378–400°F)

HEAT-DISTORTION TEMPERATURE: 138°C (280°F)

MAXIMUM CONTINUOUS SERVICE TEMPERATURE: 260°C (500°F)

MELTING/DECOMPOSITION TEMPERATURE: approx. 300°C (527°F), totally decomposes by 700°C (1,290°F)

SPECIFIC GRAVITY: 1.34

THERMAL CONDUCTIVITY: 0.25 W/mK

COEFFICIENT OF LINEAR EXPANSION: 0.000055°K (0.00003°F)

HARDNESS: 86 Shore hardness D

ELONGATION: 3%

TENSILE MODULUS: 3400 MPa (493,000 psi)

TENSILE STRENGTH: 75 MPa (10,900 psi)

IMPACT STRENGTH: 109 J/m (204 ft-lb/in) for unnotched, 22 J/m (0.4 ft-lb/in) for notched

FLEXURAL STRENGTH: 141 MPa (20,300 psi)

FLEXURAL MODULUS: 4200 MPa (609,000 psi)

COMPRESSIVE STRENGTH: 113 MPa (16,400 psi)

ABRASION RESISTANCE: good

CREEP: for an applied stress of 7 MPa (1,000 psi) the creep modulus is 13,000 MPa (1,905,000 psi) at 10 hr, 12,700 MPa (1,838,000 psi) at 1000 hr and 12,600 MPa (1,827,000 psi) at 10,000 hr

COEFFICIENT OF FRICTION: 0.24

ELECTRICAL RESISTIVITY: 10E16 Ohm/cm for volume

DIELECTRIC STRENGTH: 60 kV/mm

DIELECTRIC CONSTANT: 3.1 at 50 Hz; 3.2 at 10E6 Hz

DISSIPATION FACTOR: 0.0004 at 50 Hz; 0.0007 at 10E6 Hz

TRANSMISSION OF LIGHT: low

WATER ABSORPTION: 0.02% of volume

FLAMMABILITY: will not flame

COMBUSTION PRODUCTS: hydrogen, methane, carbon monoxide, carbon dioxide, carbon oxysulfide; gray smoke

BURNING RATE: self-extinguishing

SELF-IGNITION TEMPERATURE: 540°C (1000°F)

CHEMICAL STABILITY; RESISTANCE TO:

ACIDS: good

BASES: good

SOLVENTS: excellent

TOXICITY: suitable for food use

WEAKNESSES: difficult to process but resulting molding excellent, high cost

MONOMER:

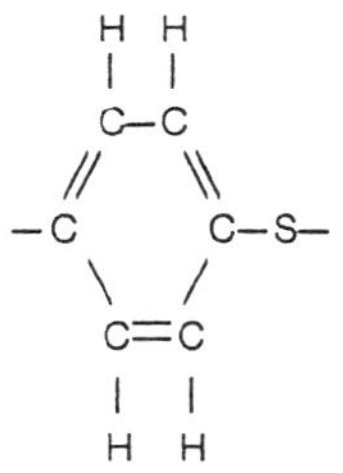

STRUCTURE: highly crystalline, can cross-link on heating

APPLICATIONS: electrical and electronic components, oil field and chemical processing equipment due to solvent resistance coupled with high temperature stability and low coefficient of friction

POLYPROPYLENE (PP)

GLASS TRANSITION TEMPERATURE: −27°C (−17°F)

HEAT-DISTORTION TEMPERATURE: 45–120°C (113–248°F) for solid and foam

MAXIMUM CONTINUOUS SERVICE TEMPERATURE: 100°C (212°F)

MINIMUM CONTINUOUS SERVICE TEMPERATURE: −30°C (−22°F)

MELTING/DECOMPOSITION TEMPERATURE: 176°C (349°F) for melting

SPECIFIC GRAVITY: 0.90–0.907

SPECIFIC HEAT: 2.0 kJ/kgK

THERMAL CONDUCTIVITY: 0.17–0.22 W/mK

COEFFICIENT OF LINEAR EXPANSION: 0.00015°K (0.00008°F) for solid and foam

HARDNESS: 36–70 ball indention N/mm²

ELONGATION: 20–800%

TENSILE MODULUS: 1100–1300 MPa (160,000–190,000 psi)

TENSILE STRENGTH: 21–39 MPa (3,100–5,700 psi); 15 MPa (2,100 psi) for 600 kg/m³ (38 lb/ft³) density foam

IMPACT STRENGTH: no break for unnotched; 3–17 J/mm notched

STIFFNESS IN TENSION: 138 MPa (200,000 psi)

FLEXURAL STRENGTH: 37 MPa (5,300 psi), 22 MPa (3,200 psi) for 600 kg/m³ (38 lb/ft³) density foam

FLEXURAL MODULUS: 2,620 MPa (380,000 psi), 827 MPa (120,000 psi) for 600 kg/m³ (38 lb/ft³) density foam

DEFLECTION TEMPERATURE UNDER FLEXURAL LOAD: 52–60°C (125–140°F) for a load of 1.8 MPa (264 psi); 93–121°C (200–250°F) for a load of 0.5 MPa (66 psi)

COMPRESSIVE STRENGTH: 38–55 MPa (5,500–8,000 psi)

ABRASION RESISTANCE: good

CREEP: under ambient conditions and an applied stress of 6.9 MPa (1,000 psi) the creep modulus is 1,138 MPa (165,000 psi) at 1 hr, 621 MPa (90,000 psi) at 100 hr and 448 MPa (65,000 psi) at 1000 hr

COEFFICIENT OF FRICTION: 0.28–0.67, 0.3 against austenitic steel

ELECTRICAL RESISTIVITY: greater than 10E17 by volume; 10E13 on the surface

DIELECTRIC STRENGTH: 50–56 kV/mm

DIELECTRIC CONSTANT: 2.27 at 50 Hz; 2.25 at 10E6 Hz

ARC RESISTANCE: 140 sec

DISSIPATION FACTOR: less than 0.0004 at 50 Hz; less than 0.0005 at 10E6 Hz

REFRACTIVE INDEX: 1.49

TRANSMISSION OF LIGHT: 60%, usually milky-white, 3% haze

WATER ABSORPTION: 0.01–0.03% of volume

GAS TRANSMISSION RATE: 187 cc of oxygen through 1 mil film/24 hrs/100 sq in at 1 atm; 639 cc of carbon dioxide under the same conditions

FLAMMABILITY: will burn with blue flame

BURNING RATE: slow

CHEMICAL STABILITY; RESISTANCE TO:
ACIDS: good
BASES: excellent
SOLVENTS: good, excellent against fuels and gasoline
OILS AND FATS: excellent
SUNLIGHT: poor, crazes
BACTERIAL AND FUNGAL ATTACK: will not support growth

TOXICITY: some food-containing grades are available

RESPONSE TO IRRADIATION: cross-links

WEAKNESSES: poor dyeability; embrittlement at low temperature, 0°C (32°F) and lower

MONOMER: 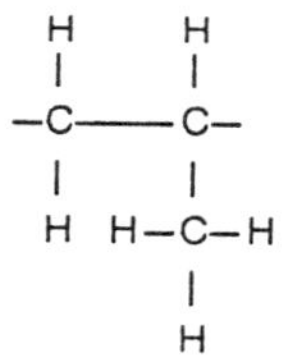

STRUCTURE: thermoplastic, crystallizes up to 75%

APPLICATIONS: pipes, chain shells, capacitor dielectrics, rope due to enduring flex-fatigue resistance

POLYSTYRENE (PS)

GLASS TRANSITION TEMPERATURE: 100°C (212°F)

HEAT-DISTORTION TEMPERATURE: 78–99°C (172–210°F)

MAXIMUM CONTINUOUS SERVICE TEMPERATURE: 50–70°C (122–158°F)

MINIMUM CONTINUOUS SERVICE TEMPERATURE: −10°C (14°F)

MELTING/DECOMPOSITION TEMPERATURE: 240°C (464°F)

SPECIFIC GRAVITY: 1.05

SPECIFIC HEAT: 1.3 kJ/kgK

THERMAL CONDUCTIVITY: 0.18 W/mK; 0.24 Btu-in/F-ft^2 for foam of density 2 lb/ft^3 this value increases both with increase and decrease of density, 0.26 Btu/h-ft^2-F/in for foam densities of both 1 lb/ft^3 and 7 lb/ft^3

COEFFICIENT OF LINEAR EXPANSION: 0.00007°K (0.00004°F); 0.0003–0.0004°F for foams in the density range 1–4 lb/ft^3 and temperature range −30 to +30°C

HARDNESS: 120–130 ball indention N/mm^2

ELONGATION: 3–4%

TENSILE MODULUS: 2750–4100 MPa (400,000–600,000 psi)

TENSILE STRENGTH: 45–65 MPa (6500–9400 psi); 1.6 MPa (225 psi) for a foam of 53 kg/m^3 (3.3 lb/ft^3) density and 0.15 mm (0.006 in) cell size; 0.4 MPa (60 psi) for a foam of 35 kg/m^3 (2.2 lb/ft^3) density and cell size of 0.4 mm (0.014 in)

IMPACT STRENGTH: 5–20 J/mm (94 ft-lb/in); 2 J/mm (37 ft-lb/in) for notched

FLEXURAL STRENGTH: 96 MPa (14,000 psi); 0.6 MPa (85 psi) for a foam of 53 kg/m^3 (3.3 lb/ft^3) density and cell size of 0.15 mm (0.006 in); 1 MPa (140 psi) for a foam of density 35 kg/m^3 (2.2 lb/ft^3) and 0.4 mm (0.014 in) cell size

FLEXURAL MODULUS: 3300 MPa (480,000 psi)

SHEAR STRENGTH: 0.5 MPa (65 psi) for a foam of 53 kg/m^3 (3.3 lb/ft^3) density and cell size of 0.15 mm (0.006 in); 0.24 MPa (35 psi) for a foam of 35 kg/m^3 (2.2 lb/ft^3) density and a cell size of 0.4 mm (0.014 in)

SHEAR MODULUS: 6 MPa (875 psi) for a foam of density 53 kg/m^3 (3.3 lb/ft^3) and cell size of 0.15 mm (0.006 in); 8.3 MPa (1200 psi) for a foam of density 35 kg/m^3 (2.2 lb/ft^3) and cell size of 0.4 mm (0.014 in)

COMPRESSIVE STRENGTH: 0.83 MPa (120 psi) for a foam of density 53 kg/m^3 (3.3 lb/ft^3) and cell size of 0.15 mm (0.006 in); 0.21 MPa (30 psi) for a foam of density 35 kg/m^3 (2.2 lb/ft^3) and cell size of 0.4 mm (0.014 in)

ABRASION RESISTANCE: poor

CREEP: under ambient conditions and with an applied stress of 10.3 MPa (1500 psi) the creep modulus is 931 MPa (135,000 psi) after 1 hr, 138 MPa (20,000 psi) after 100 hr and 69 MPa (10,000 psi) after 1,000 hrs

COEFFICIENT OF FRICTION: 0.4–0.5, 0.46 against austenitic steel

ELECTRICAL RESISTIVITY: 10E16 for volume, 10E13 for surface; 3.8X10E13 Ohm-cm for foam volume, 9X10E13 Ohm for foam surface both of density 32 kg/m^3 (2 lb/ft^3)

DIELECTRIC STRENGTH: 300–700 KV/cm; 49 V/mil for foam of 32 kg/m^3 (2 lb/ft^3) density

DIELECTRIC CONSTANT: 2.5 at 50 Hz and 10E6 Hz; 1.0 for 10E3 to 10E9 Hz for foam of density 32 kg/m^3 (2 lb/ft^3)

ARC RESISTANCE: 60 sec

DISSIPATION FACTOR: 0.0001–0.0004 at 50 Hz and 10E6 Hz; 0.0001 for 10E3 to 10E9 Hz for foam of density 32 kg/m^3 (2 lb/ft^3)

REFRACTIVE INDEX: 1.59

TRANSMISSION OF LIGHT: 88%, 1.5% haze

MOISTURE-VAPOR TRANSMISSION: 120 g/mil/100 in^2-24 hr

WATER ABSORPTION: 0.03–0.1% of volume; similar for closed cell foam

GAS TRANSMISSION RATE: 310 cc of oxygen/24 hrs/100 in^2/atm; 1,535 cc of carbon dioxide under the same conditions

FLAMMABILITY: burns with yellow flame, heat of combustion is 16,000 Btu/lb

COMBUSTION PRODUCTS: carbon dioxide, carbon monoxide, water and carbon in the form of soot

FLASH IGNITION TEMPERATURE: 360°C (680°F)

LIMITING OXYGEN FOR SUSTAINED IGNITION: 18%

BURNING RATE: slow, 0.64 mm/s (1.5 in/min)

CHEMICAL STABILITY; RESISTANCE TO:

- *ACIDS:* good except for strong oxidizing acids like nitric and sulfuric, chlorine and bromine which react with both solid and foam
- *BASES:* excellent for solid and foam form
- *SOLVENTS:* sensitive to esters, ketones and heavier alcohols as well as gasoline; the foams are even more affected due to larger surface area
- *OILS AND FATS:* excellent
- *SUNLIGHT:* exposure causes discoloration which is more rapid in foams especially if they have color added
- *BACTERIAL AND FUNGAL ATTACK:* will not support growth for either polymer or foam form

TOXICITY: acceptable for food and personal contact unless undesirable additives have been used, almost all grades of fire retardant PS are unacceptable for food contact

RESPONSE TO IRRADIATION: cross-links

WEAKNESSES: brittle

MONOMER:

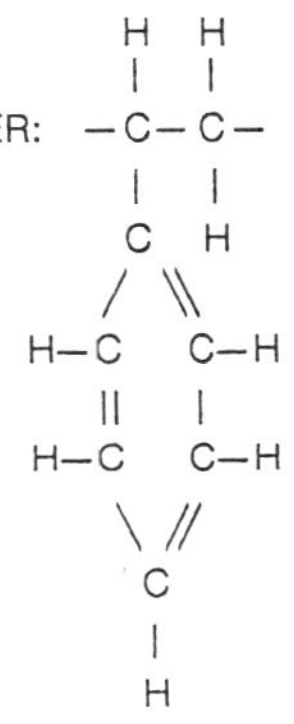

STRUCTURE: amorphous due to the large pendant, thermoplastic

APPLICATIONS: packaging, toys, building insulation; areas that require dimensional stability, low cost and excellent surface finish

POLYSULPHONE (PSU)

GLASS TRANSITION TEMPERATURE: 180 to 220°C (356–428°F)

HEAT-DISTORTION TEMPERATURE: 174°C (345°F)

MAXIMUM CONTINUOUS SERVICE TEMPERATURE: 150°C (302°F)

MINIMUM CONTINUOUS SERVICE TEMPERATURE: − 100°C (− 148°F)

SPECIFIC GRAVITY: 1.24

SPECIFIC HEAT: 1.3 kJ/kgK

THERMAL CONDUCTIVITY: 0.28 W/mK

COEFFICIENT OF LINEAR EXPANSION: 0.000054°K (0.00003°F)

HARDNESS: M69 (R120) Rockwell

ELONGATION: 25–30%

MODULUS OF ELASTICITY: 2600–2750 MPa (377,000–400,000 psi)

TENSILE STRENGTH: 50–100 MPa (7,300–14,500 psi)

IMPACT STRENGTH: 69 J/m (1.3 ft-lb/in) notched

FLEXURAL STRENGTH: 106 MPa (15,400 psi)

FLEXURAL MODULUS: 2,700 MPa (390,000 psi)

COMPRESSIVE STRENGTH: 123 MPa (17,900 psi)

ABRASION RESISTANCE: poor

CREEP: under ambient conditions and an applied stress of 28 MPa (4,000 psi) the creep modulus is 2,413 MPa (350,000 psi) at 1 hr, 2,200 MPa (320,000 psi) at 100 hr and 2,137 MPa (310,000 psi) at 1000 hr

ELECTRICAL RESISTIVITY: 10E16 Ohm/cm for volume, 10E16 Ohm for surface

DIELECTRIC STRENGTH: 42 kV/mm

DIELECTRIC CONSTANT: 3.1 at 50 Hz; 3.0 at 10E6 Hz

ARC RESISTANCE: 122 sec

DISSIPATION FACTOR: 0.0008 at 50 Hz; 0.003 at 10E6 Hz

REFRACTIVE INDEX: 1.63

TRANSMISSION OF LIGHT: high, 5% haze

WATER ABSORPTION: 0.02% of volume

GAS PERMEABILITY: very low

FLAMMABILITY: ignites

COMBUSTION PRODUCTS: black smoke

BURNING RATE: very low; self-extinguishing

CHEMICAL STABILITY; RESISTANCE TO:

- *ACIDS:* excellent
- *BASES:* excellent
- *SOLVENTS:* poor
- *OILS AND FATS:* good
- *SUNLIGHT:* excellent
- *BACTERIAL AND FUNGAL ATTACK:* good

TOXICITY: used in sterilizable medical instruments

RESPONSE TO IRRADIATION: excellent resistance to gamma radiation

WEAKNESSES: high price

MONOMER:

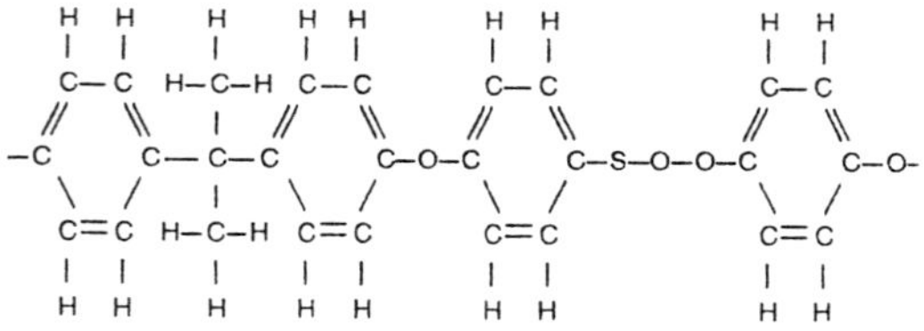

STRUCTURE: a family of thermoplastics that do not crystallize and can be used as elastomers

APPLICATIONS: components for high temperature electrical use; food processing equipment due to hydrolytic and heat stability; hoses due to excellent oil and gasoline resistance

POLYTETRAFLUOROETHYLENE, teflon (PTFE)

GLASS TRANSITION TEMPERATURE: −97°C (−143°F)

HEAT-DISTORTION TEMPERATURE: 115–125°C (239–257°F)

MAXIMUM CONTINUOUS SERVICE TEMPERATURE: 250°C (432°F)

MINIMUM CONTINUOUS SERVICE TEMPERATURE: −200°C (−328°F)

MELTING/DECOMPOSITION TEMPERATURE: 327°C (621°F) for melting

SPECIFIC GRAVITY: 2.15–2.20

SPECIFIC HEAT: 1.0 KJ/kgK

THERMAL CONDUCTIVITY: 0.25 W/mK

COEFFICIENT OF LINEAR EXPANSION: 0.0001°K (0.00006°F)

HARDNESS: 27–35 ball indention N/mm^2

ELONGATION: 350–550%; 13% for fiber form

TENSILE MODULUS: 410 MPa (60,000 psi) this increases with pressure until a plateau at 550 MPa (80,000 psi) and 1,860 MPa (270,000 psi) modulus is reached, then the modulus remains constant with further increases in pressure

TENSILE STRENGTH: 25–36 MPa (3,600–5,200 psi); 345 MPa (50,000 psi) for fiber form

EFFECT OF PRESSURE ON TENSILE STRENGTH: under 207 MPa (30,000 psi) pressure the rupture strain is 0.6 and the stress is 46 MPa (6,600 psi); under 414 MPa (60,000 psi) pressure the rupture strain is 0.3 and the stress is 62 MPa (9,000 psi)

IMPACT STRENGTH: no break for unnotched; 13–15 J/mm for notched

FLEXURAL MODULUS: 345 MPa (50,000 psi) at 50% crystallinity, 1,517 MPa (220,000 psi) at 95% crystallinity

DEFLECTION TEMPERATURE UNDER FLEXURAL LOAD: 56°C (132°F) under a load of 1.8 MPa (264 psi); 121°C (250°F) under a load of 0.5 MPa (66 psi)

COMPRESSIVE STRENGTH: 12 MPa (1700 psi)

ABRASION RESISTANCE: poor

COEFFICIENT OF FRICTION: 0.02–0.04, 0.11 against austenitic steel

CREEP: under ambient conditions and an applied stress of 3.5 MPa (500 psi) the creep modulus is 345 MPa (50,000 psi) at 1 hr, 207 MPa (30,000 psi) at 100 hr and 172 MPa (25,000 psi) at 1000 hr; for an applied stress of 6.9 MPa (1,000 psi) the creep modulus is 138 MPa (20,000 psi) at 1 hr, 70 MPa (10,000 psi) at 100 hr and 35 MPa (5,000 psi) at 1000 hr

COEFFICIENT OF FRICTION: 0.04–0.05

ELECTRICAL RESISTIVITY: 10E18 Ohm/cm for volume; 10E17 Ohm for surface

DIELECTRIC STRENGTH: 48 KV/mm

DIELECTRIC CONSTANT: 2.1 at 50 Hz; 2.1 at 60 Hz and 10E6 Hz

ARC RESISTANCE: 300 sec

DISSIPATION FACTOR: 0.0002 at 50 Hz and 10E6 Hz

REFRACTIVE INDEX: 1.35

TRANSMISSION OF LIGHT: low

MOISTURE-VAPOR TRANSMISSION: low; 3 g/mil/100 in^2-24 hr

WATER ABSORPTION: 0% of volume

GAS TRANSMISSION RATE: 1100 cc of oxygen through 1 mil/24 hrs/100 in^2/atm; 3000 cc of carbon dioxide under the same conditions

FLAMMABILITY: burns

COMBUSTION PRODUCTS: fumes may be toxic

BURNING RATE: self-extinguishing

SELF-IGNITION TEMPERATURE: 530°C (986°F)

CHEMICAL STABILITY; RESISTANCE TO:
- *ACIDS:* excellent
- *BASES:* excellent
- *SOLVENTS:* excellent; insoluble in all organics
- *OILS AND FATS:* excellent
- *SUNLIGHT:* excellent

RESPONSE TO IRRADIATION: degradation due to breaking of C–C bonds

WEAKNESSES: high melt viscosity so cannot be processed by conventional molding

MONOMER: 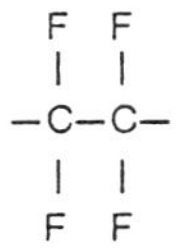

STRUCTURE: thermoplastic that can crystallize up to 95%. In the crystal phase, there is a first-order transformation at 20°C and a second-order transformation at 30°C

APPLICATIONS: due to corrosion resistance, exceptionally low coefficient of friction and good toughness, PTFE is used for bearings. Its outstanding electrical properties make it useful for high-frequency, high-temperature cable insulation.

POLY(UREA-FORMALDEHYDE), urea (UF)

GLASS TRANSITION TEMPERATURE: decomposes first

HEAT-DISTORTION TEMPERATURE: 130°C (266°F)

MAXIMUM CONTINUOUS SERVICE TEMPERATURE: 70°C (158°F)

SPECIFIC GRAVITY: 1.5

SPECIFIC HEAT: 1.2 kJ/kgK

THERMAL CONDUCTIVITY: 0.4 W/mK

COEFFICIENT OF LINEAR EXPANSION: 0.00005°K (0.00003°F)

HARDNESS: 350 ball indention N/mm²

ELONGATION: 1%

TENSILE MODULUS: 10,500 MPa (1,500,000 psi)

TENSILE STRENGTH: 30 MPa (4,400 psi)

IMPACT STRENGTH: 6.5 J/mm (122 ft-lb/in) for unnotched; 1.5 J/mm (28 ft-lb/in) for notched

FLEXURAL STRENGTH: 103 MPa (15,000 psi)

FLEXURAL MODULUS: 8,960 MPa (1,300,000 psi)

COMPRESSIVE STRENGTH: 207 MPa (30,000 psi)

ABRASION RESISTANCE: good

ELECTRICAL RESISTIVITY: 10E11 Ohm/cm for volume; 10E10 Ohm for surface

DIELECTRIC STRENGTH: 30–40 kV/mm

DIELECTRIC CONSTANT: 8 for 50 Hz; 7 for 10E6 Hz

ARC RESISTANCE: 120 sec

DISSIPATION FACTOR: 0.04 for 50 Hz; 0.3 for 10E6 Hz

REFRACTIVE INDEX: 1.54–1.56

TRANSMISSION OF LIGHT: 20%

WATER ABSORPTION: 0.4–0.8% of volume

BURNING RATE: self-extinguishing

CHEMICAL STABILITY; RESISTANCE TO:
ACIDS: good
BASES: good
SOLVENTS: good to excellent
OILS AND FATS: excellent
SUNLIGHT: fades

RESPONSE TO IRRADIATION: cross-links

WEAKNESSES: fair machining qualities

MONOMER: 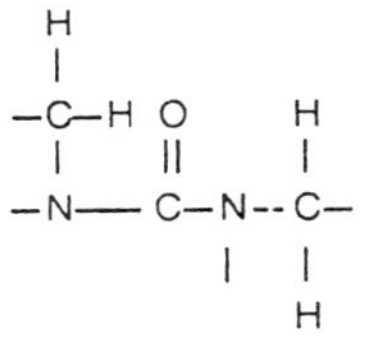

STRUCTURE: cross-linked thermoset which does not crystallize

APPLICATIONS: low cost gives many applications, among them: buttons, electrical fittings, and toilet seats

POLYURETHANE (PUR)

HEAT-DISTORTION TEMPERATURE: 127–143°C (206–290°F)

MAXIMUM CONTINUOUS SERVICE TEMPERATURE: 80°C (176°F)

MINIMUM CONTINUOUS SERVICE TEMPERATURE: −40°C (−40%F)

SPECIFIC GRAVITY: 1.2

SPECIFIC HEAT: 0.5 KJ/kgK

THERMAL CONDUCTIVITY: 1.7 W/mK

HARDNESS: M110–M120 Rockwell

ELONGATION: 400–450%; 100–340% for foam

TENSILE MODULUS: 700 MPa (102,000 psi)

TENSILE STRENGTH: 30–40 MPa (4,350–5,800 psi); 0.5 MPa (70 psi) for foam of density 40 kg/m³ (2.5 lb/ft³)

IMPACT STRENGTH: no break in notched and unnotched tests

FLEXURAL STRENGTH: 0.4 MPa (60 psi) for a foam of density 40 kg/m³ (2.5 lb/ft³)

COMPRESSIVE STRENGTH: 0.3 MPa (40 psi) for a foam of density 40 kg/m³ (2.5 lb/ft³)

ABRASION RESISTANCE: very good

ELECTRICAL RESISTIVITY: 10E12 Ohm/cm for volume; 10E11 Ohm for surface

DIELECTRIC STRENGTH: 30–60 kV/mm

DIELECTRIC CONSTANT: 6.5 at 50 Hz, 5.6 at 10E6 Hz; 1.02–1.5 at 10E6 Hz for foam

DISSIPATION FACTOR: 0.03 at 50 Hz, 0.06 at 10E6 Hz; 0.0005–0.005 at 10E5 Hz for foam

TRANSMISSION OF LIGHT: clear to opaque

SOUND ABSORPTION: good for open-celled foams

WATER ABSORPTION: 0.7–0.9% of volume

FLAMMABILITY: moderate, burns with a yellow flame

COMBUSTION PRODUCTS: black smoke

CHEMICAL STABILITY; RESISTANCE TO:
ACIDS: good for weak acids but will not withstand strong ones
BASES: same as for acids
SOLVENTS: susceptible to aromatic solvents such as methyl ethyl ketone
OILS AND FATS: good resistance for most grades
SUNLIGHT: poor
BACTERIAL AND FUNGAL ATTACK: good resistance for most grades except ester-linked PUR which supports growth
OZONE: resists cracking in the presence of ozone much better than most other elastomers

TOXICITY: most foam grades toxic

RESPONSE TO IRRADIATION: cross-linking at the nitrogen atoms

WEAKNESSES: more expensive than most other polymers; cannot withstand exposure to steam, susceptible to static build-up

MONOMER:

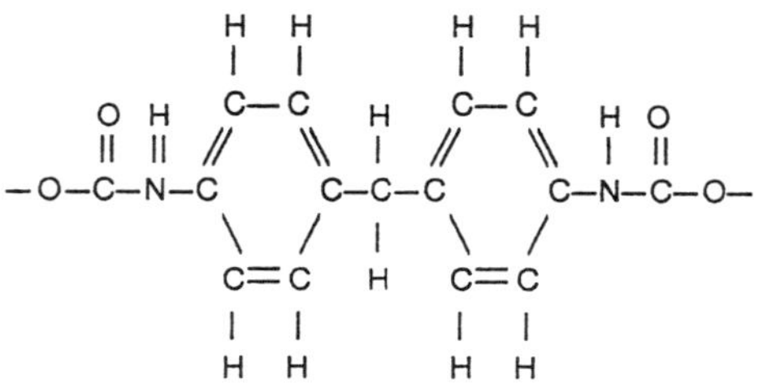

STRUCTURE: PUR is a family of polymers which vary from soft linear thermoplastic elastomers to hard cross-linked, thermoset rigid foams. The values here are for the slightly cross-linked thermoset elastomer except in the foam section.

APPLICATIONS: mining operations due to toughness, tear resistance and strength; paper production; roller skate wheels; forklift tires; hammers; meat-processing due to Federal Food and Drug approval. Foam finds a wide market in both the flexible and rigid form. Construction uses have increased dramatically for the rigid form, while flooring and bedding lead the uses of the flexible form.

POLY(VINYL CHLORIDE), vinyl (PVC)

GLASS TRANSITION TEMPERATURE: 80°C (176°F)

HEAT-DISTORTION TEMPERATURE: 75–110°C (167–230°F)

MAXIMUM CONTINUOUS SERVICE TEMPERATURE: 65–85°C (149–185°F)

MINIMUM CONTINUOUS SERVICE TEMPERATURE: −5°C (23°F)

MELTING/DECOMPOSITION TEMPERATURE: approx. 185°C (365°F) melting

SPECIFIC GRAVITY: 1.38–1.55

SPECIFIC HEAT: 0.85–0.9 KJ/kgK

THERMAL CONDUCTIVITY: 0.14–0.17 W/mK

COEFFICIENT OF LINEAR EXPANSION: 0.00007–0.00008°K (0.00004°F)

HARDNESS: 75–150 ball indention N/mm²

ELONGATION: 10–50%; 41–20% for fiber form

TENSILE MODULUS: 1000–3500 MPa (145,000–510,000 psi)

TENSILE STRENGTH: 50–75 MPa; 330–370 MPa (48,000–54,000 psi) for fiber

IMPACT STRENGTH: 20 J/mm—no break for unnotched; 2–50 J/mm for notched

FLEXURAL STRENGTH: 83 MPa (12,000 psi)

FLEXURAL MODULUS: 3,500 MPa (500,000 psi)

COMPRESSIVE STRENGTH: 62 MPa (9,000 psi)

ABRASION RESISTANCE: very good

CREEP: under ambient conditions and an applied stress of 6.9 MPa (1,000 psi) the creep modulus is 2,275 MPa (330,000 psi) after 1 hr, 1,930 MPa (280,000 psi) after 100 hr, and 1,240 MPa (180,000 psi) after 1,000 hr

COEFFICIENT OF FRICTION: 0.4–0.45, 0.6 against austenitic steel

ELECTRICAL RESISTIVITY: 10E15 Ohm/cm for volume; 10E13 Ohm for surface

DIELECTRIC STRENGTH: 35–50 KV/mm

DIELECTRIC CONSTANT: 3.5 at 50 Hz; 3 at 60 Hz; 3.0 at 10E6 Hz

ARC RESISTANCE: 70 sec

DISSIPATION FACTOR: 0.011 at 50 Hz, 0.015 at 10E6 Hz

REFRACTIVE INDEX: 1.52–1.55

TRANSMISSION OF LIGHT: variable

WATER ABSORPTION: 0.04–0.4% of volume

GAS PERMEABILITY: low, 23-32X10E17 cm³s/g for oxygen 40-180X10E17 cm³s/g for carbon dioxide

FLAMMABILITY: burns with a yellow flame

COMBUSTION PRODUCTS: sharp acrid fumes of HCl, carbon monoxide, carbon dioxide

FLASH IGNITION TEMPERATURE: 454°C (850°F)

LIMITING OXYGEN FOR SUSTAINED IGNITION: 47%

BURNING RATE: self-extinguishing

CHEMICAL STABILITY; RESISTANCE TO:
- *ACIDS:* excellent
- *BASES:* good
- *SOLVENTS:* good except ethers and esters
- *OILS AND FATS:* excellent
- *SUNLIGHT:* excellent
- *BACTERIAL AND FUNGAL ATTACK:* will support growth if plasticized

TOXICITY: most grades approved for use with foods

RESPONSE TO IRRADIATION: for doses up to 8 megarads it cross-links

WEAKNESSES: unplasticized PVC is difficult to process

MONOMER: 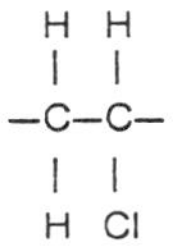

STRUCTURE: thermoplastic, maximum crystallinity is 10%

APPLICATIONS: pipes and pipe fittings, rain gutters, automotive interior parts; wire and cable insulation; flooring; nearly any application that requires dimensional stability and low cost

POLY(VINYLIDENE CHLORIDE), saran (VC)

GLASS TRANSITION TEMPERATURE: −18°C (0°F)

HEAT-DISTORTION TEMPERATURE: 45–55°C (113–131°F)

MAXIMUM CONTINUOUS SERVICE TEMPERATURE: 160–200°F

MELTING/DECOMPOSITION TEMPERATURE: 210°C (410°F) melting

SPECIFIC GRAVITY: 1.65–1.72

SPECIFIC HEAT: 0.32

THERMAL CONDUCTIVITY: 0.126 W/mK

COEFFICIENT OF LINEAR EXPANSION: 0.00019°K (0.00011°F)

HARDNESS: M50–M65 Rockwell

ELONGATION: 250%

TENSILE MODULUS: 345–550 MPa (50,000–80,000 psi)

TENSILE STRENGTH: 21–35 MPa (3,000–5,000 psi)

IMPACT STRENGTH: 16–53 J/m (0.3–1.0 ft-lb/in) for notched

FLEXURAL STRENGTH: 29–43 MPa (4,200–6,200 psi)

COMPRESSIVE STRENGTH: 14–19 MPa (2,000–2,700 psi)

ABRASION RESISTANCE: good

CREEP: high

ELECTRICAL RESISTIVITY: 10E14–10E16 Ohm/cm for volume

DIELECTRIC STRENGTH: 15–24 kV/mm

DIELECTRIC CONSTANT: 3–4 at 10E6 Hz

DISSIPATION FACTOR: 0.05–0.08 at 10E6 Hz

REFRACTIVE INDEX: 1.60–1.63

TRANSMISSION OF LIGHT: clear to milky-white

MOISTURE-VAPOR TRANSMISSION: low; 1 g/mil/100 in^2-24 hr

WATER ABSORPTION: low % of volume

GAS PERMEABILITY: 3.6X10E17 cm^3s/g for oxygen, 14-23X10E17 cm^3s/g for carbon dioxide

FLAMMABILITY: ignites

COMBUSTION PRODUCTS: sharp acrid fumes

BURNING RATE: self-extinguishing

FLASH IGNITION TEMPERATURE: 532°C (990°F)

CHEMICAL STABILITY; RESISTANCE TO:
- *ACIDS:* very good
- *BASES:* very good
- *SOLVENTS:* very good

TOXICITY: most grades acceptable for food use

RESPONSE TO IRRADIATION: degrades

MONOMER: 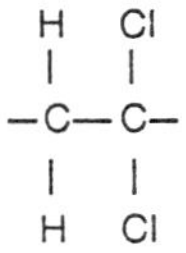

STRUCTURE: highly crystalline and generally insoluble

APPLICATIONS: film and sheeting for food packaging; pipes and tapes

SILICONE RUBBER (SI)

GLASS TRANSITION TEMPERATURE: −130°C (−202°F)

MAXIMUM CONTINIUOUS SERVICE TEMPERATURE: 250°C (482°F)

MINIMUM CONTINUOUS SERVICE TEMPERATURE:

−30°C (−22°F) due to onset of crystallization

MELTING/DECOMPOSITION TEMPERATURE: −58°C (−72°F) above this SI goes from a true solid to a viscous liquid

SPECIFIC GRAVITY: 1.2

SPECIFIC HEAT: 0.8 kJ/kgK

THERMAL CONDUCTIVITY: 0.0007 W/mK

COEFFICIENT OF LINEAR EXPANSION: 0.00012°K (0.00007°F)

HARDNESS: 45 Ball indention N/mm²

ELONGATION: 200%

TENSILE STRENGTH: 3.45 MPa (500 psi); retains its strength in working environment up to 150°C (302°F)

COMPRESSIVE STRENGTH: 103 MPa (15,000 psi)

IMPACT STRENGTH: no break for both notched and unnotched

FLEXURAL STRENGTH: retains flexibility in working environment down to −90°C (−130°F); silicones have the highest backbone flexibility of any chain

ABRASION RESISTANCE: poor

ELECTRICAL RESISTIVITY: 10E14 ohm/cm for volume; 10E12 for surface

DIELECTRIC STRENGTH: 20–40 kV/mm

DIELECTRIC CONSTANT: 4 at 50 Hz; 3.5 at 10E6 Hz

MOISTURE-VAPOR TRANSMISSION: 170 g/mil/100 in²-24 hr

WATER ABSORPTION: 0.2% of volume

GAS TRANSMISSION RATE: 98,000 cc of oxygen can be transmitted through a 1 mil film in 24 hours over a 100 square inch area with one atmosphere of differential pressure. 519,000 cc of carbon dioxide can be transmitted under the same conditions

FLAMMABILITY: will not burn

COMBUSTION PRODUCTS: silica producing white smoke

CHEMICAL STABILITY; RESISTANCE TO:

ACIDS: fair

BASES: excellent

SOLVENTS: poor

OXYGEN: less susceptible than carbon-based polymers

SUNLIGHT: excellent

BACTERIAL AND FUNGAL ATTACK: excellent

TOXICITY: none for medical grades

RESPONSE TO IRRADIATION: cross-links for doses up to 5 megarads

WEAKNESSES: rigid processing requirements make it fairly expensive

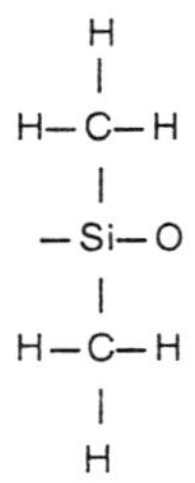

STRUCTURE: the replacement of a group such as CH_2 with a single atom invariably results in a more flexible chain. Thus the alternating of the oxygen atom in the silicone backbone insures a flexible chain, since the oxygen cannot bond beyond the continuation of the chain. Unvulcanized silicone is used in its grease form as a lubricant. Vulcanization by internal cross-linking is usually limited to about one link in 200 chain units. Cross-linking is also done by vulcanizing agents.

APPLICATIONS: gaskets and seals in aircraft, and heating equipment due to the wide range of application temperature; electrical insulation under water; calking and glazing; many medical uses such as artificial heart valves and replacement sections for arteries

UNSATURATED POLYESTER RUBBER and ALKYD (UP, PER, PEA)

GLASS TRANSITION TEMPERATURE: 100–170°C (212–338°F)

MAXIMUM CONTINUOUS SERVICE TEMPERATURE: 150°C (302°F)

SPECIFIC GRAVITY: 2.0 for resin; 1.10–1.46 for rubber and alkyd

SPECIFIC HEAT: 1.2 kJ/kgK

THERMAL CONDUCTIVITY: 0.7 W/mK

COEFFICIENT OF LINEAR EXPANSION: 0.000002–0.000004°K (0.000001–0.000002°F)

HARDNESS: 240 Ball indention N/mm² for resin; 15A for rubber

ELONGATION: 300% for rubber

TENSILE MODULUS: 2,070–4,140 MPa (300,000–600,000 psi) for alkyd

TENSILE STRENGTH: 41–90 MPa (6,000–13,000 psi) for rubber and alkyd

IMPACT STRENGTH: 4.5 J/mm for unnotched; 3.0 J/mm for notched

FLEXURAL STRENGTH: 115 MPa (16,700 psi)

FLEXURAL MODULUS: 2,760 MPa (400,000 psi); 517 MPa (75,000 psi) for rubber

COMPRESSIVE STRENGTH: 128 MPa (18,600 psi)

BURNING RATE: 0.62 mm/s

CHEMICAL STABILITY; RESISTANCE TO:

ACIDS: good except in concentrated form for resin and alkyd fair for rubber

BASES: poor for resin and alkyd; fair for rubber

SOLVENTS: good

OILS AND FATS: excellent

SUNLIGHT: excellent for resin; good for alkyd

BACTERIAL AND FUNGAL ATTACK: alkyds susceptible to attack

RESPONSE TO IRRADIATION: cross-links

WEAKNESSES: poor machining qualities; subject to static build-up

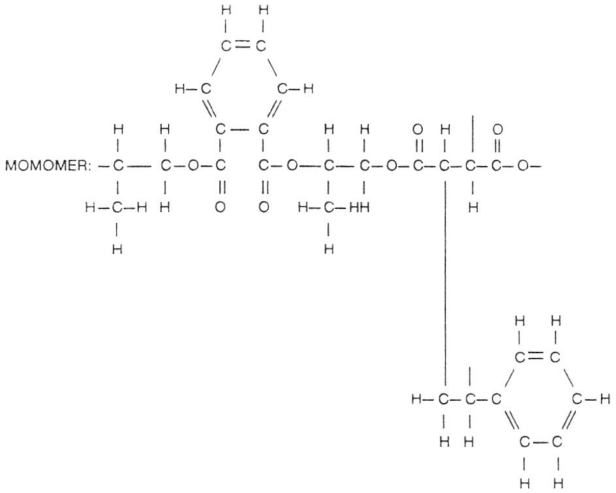

STRUCTURE: cross-linked family of thermosets which do not crystallize; can be made into a rubber; under some circumstances it can also be a thermoplastic

APPLICATIONS: in the reinforced state, it is used extensively for boat hulls, appliances and shower stalls

ABRASION RESISTANCE: very good, excellent for rubber

CREEP: under ambient conditions and an applied stress of 6.9 MPa (1,000 psi) the creep modulus is 3,033 MPa (440,000 psi) at 1 hr, 2,760 MPa (400,000 psi) at 100 hrs and 2,620 MPa (380,000 psi) at 1000 hrs for the resin; for an applied stress if 21 MPa (3,000 psi) the creep modulus is 18,620 MPa (2,700,000 psi) at 1 hr and 15,860 MPa (2,300,000 psi) at 100 hrs for alkyd

ELECTRICAL RESISTIVITY: 10E12 Ohm-cm for volume; 10E10 Ohm for surface

DIELECTRIC STRENGTH: 25–53 kV/mm

DIELECTRIC CONSTANT: 6 at 50 Hz; 5 at 10E6 Hz

ARC RESISTANCE: 170 sec

DISSIPATION FACTOR: 0.04 at 50 Hz; 0.02 at 10E6 Hz

WATER ABSORPTION: 0.03–0.5% of volume

GAS PERMEABILITY: 11 cc of oxygen through 1 mil film/24 hrs/100 in² at 1 atm; 19 cc of carbon dioxide under the same conditions

FLAMMABILITY: burns with a yellow flame

COMBUSTION PRODUCTS: carbon dioxide, carbon monoxide

FLASH IGNITION TEMPERATURE: 485°C (905°F)

GLOSSARY

The glossary terms are defined on the basis of only polymeric use, although the word may have a more general meaning.

Ablation The controlled destruction of a thermosetting polymer. This almost always occurs under conditions of high temperature, causing the surface of the polymer to char. The charring process absorbs a great deal of energy, and the nearly pure carbon surface that remains restricts heat flow to the interior of the polymer.

Abrasion The removal of material by the cutting action of a second (abrading) material or particle. Polymers are soft and subject to abrasion by most other materials including dust.

Absorption (of light) That fraction of light that fails to emerge from a polymer with the main beam of transmitted light is considered to be absorbed by the polymer. Absorption includes light that has been reflected, refracted and converted internally into heat. Absorption is very slight for thin sections of pure polymer but increases considerably with the inclusion of voids and additives.

Addition A polymerizing reaction that results in a product incapable of further reaction. Also see *condensation.*

Additive Any material added to a polymer such that it is not directly ionically or covalently bonded to the polymer. When additive levels are below approximately 20% by weight, they modify mechanical behavior without usually imparting their own characteristics. At higher additive concentrations, the resulting material can be considered a composite rather than a modified polymer.

Agglomeration The coalescing of small particles into a clump. This is particularly a problem in the processing of rough, irregularly shaped ceramic particles.

Aging (re: plasticizers) Degradation of a plasticizer prior to the end of the useful life of the polymer that contains it.

Anisotropic Having varying property values with respect to direction. In a polymer this can best be shown by considering the strength down a polymeric chain as opposed to the strength between polymeric chains.

Annealing The process by which stresses are relieved in a structure through the addition of heat. The added heat causes the material to have more micro-mobility and allows for rearrangement on a segmental basis to a lower energy structure, which is usually a more ordered array. For a crystal, this means a reduction in crystal defects.

Antioxidants Additives that inhibit a polymer from chemically reacting with oxygen. An oxygen reaction usually results in chain scission and, therefore, possible cross-linking. Oxygen reactions can also produce volatile products that contribute to further combustion of the polymer. Antioxidants are particularly useful both at combustion control and at elevated temperatures where the oxygen-polymer reaction proceeds at an accelerated rate.

Arc Resistance The breakdown of the insulating properties of a polymer during an electrical dissipation. Usually, the electrical conduction that characterizes arc resistance is a result of a carbonized path across the surface of the polymer. This path shows itself both in the form of char and also through a visible arc of electricity.

Atactic The side groups attached to a polymer chain backbone have no regular pattern. An example is:

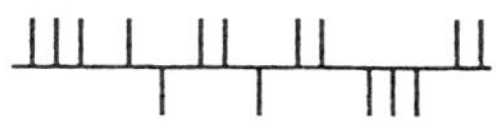

where the horizontal line represents the *backbone*

and the vertical lines represent the side groups.

Average Molecular Weight The weight average of a polymer derived by adding up the total molecular weights of all of the polymeric molecules and dividing by the total number of molecules. This does not give any indication of the range of molecular weights present in the polymer and thus has limited usefulness by itself.

Backbone The series of atoms chemically bonded in a linear array that characterizes the structure of a polymer. Most polymers are classified by their composition of backbone atoms.

Biodegradation Reduction of a polymer to its elemental form by the action of microorganisms. Only organic based polymers are subject to biodegradation. The more closely a polymer resembles its original organic structure, the more it is susceptible to this.

Blowing Agent See *Foaming Agent.*

Branched Molecules When a side group or chain has the same makeup as the main backbone of a polymer then it is said to be branched. The branch can be just one atom or as long as the main chain. Branching is often caused by the presence of multiple reaction sites during polymerization. The presence of branches hinders crystallization and may have great influence on other properties.

Brittle-Ductile Transition The temperature at which a polymer changes from brittle tensile behavior to ductile tensile behavior. This is usually not a specific temperature but rather a range of temperature on the order of 10°C. On cooling of a polymer, the ductile-brittle transition temperature may not necessarily correspond to the brittle-ductile transition temperature observed on heating of that polymer. There have been many attempts to correlate brittle-ductile transition temperature to glass transition temperature, but no reliable relationship has been found.

Brittleness A general term indicating that a polymer absorbs relatively little energy during fracture; it may simply mean that subjectively it breaks easily.

Carbon-Black A reinforcing additive used in carbon based elastomers. The presence of carbon-black in a non-crystallizing elastomer can increase the strength of the elastomer by a factor of ten. Reinforcing ability is a factor of both loading and particle size. Maximum strength is usually obtained when the loading is about 20% by volume. The smaller the particle size, the better the reinforcement to the practical limit of carbon-black. The projected optimum particle size is just smaller than the distance between cross-linking points within the elastomer. This allows for incorporation of the carbon-black without straining or breaking chains and also allows for close proximity of carbon-black particles. The maximum reinforcement seems to occur when the carbon-black forms an interlacing lattice with the elastomer. The extremely high strength of this arrangement is only in evidence at stress levels that do not disturb the aggregation of the carbon-black. Once the aggregation is broken, the extra strength associated with it is not recoverable. Carbon-black only has efficient reinforcing abilities as long as it is an amorphous material. If it is heat-treated so that it is transformed into graphite, the carbon becomes more highly bonded internally and less able to form bonds with the elastomer.

Carbon Ring A ring of six carbon atoms with three resonating double bonds. The carbon atoms are usually accompanied by a hydrogen atom. Figure 1.9 (p.7) illustrates this.

Cellular Plastic See *foam.*

Ceraplast A reinforcement ceramic particle coated with an elastomer. This arrangement allows this additive to be bonded to the elastomer, such that it reinforces via van der Waals bonds which are stronger between elastomers than between elastomer and ceramic. Since the ceramic is encased, it moves with the reinforced elastomer without tearing it.

Chain Segment The section of polymer backbone that moves together as a unit. The number of atoms which comprise a chain segment is both a function of the structure of the polymer and the temperature. The motion of chain segments allows for polymer rearrangement and stress reduction.

Chemical Bonding All matter is attracted to itself by mass attraction. In addition to mass attraction there is also the possibility of an exchange of electrons that further enhances the attraction. Usually this exchange of electrons is called chemical bonding. However, mass attraction alone is still considered a bond of the chemical type if it is between atoms in close proximity and high density, such as metal-metal bonding. Polymers are considered to be chemically bound only if they have an exchange of electrons. All other forms of attraction in polymers are considered to be of the secondary or van der Waals type of bonding. These definitions are based mainly on the strength of the bond rather than precisely on its nature.

Chemiluminescence The emission of light as a result of chemical reaction. It has proven to be a useful attribute in tracing chemical reactions during the processing of polymers as well as a means of monitoring structural changes due to testing.

cis Angular relationship to the backbone of 180 degrees. If four carbon atoms are considered to be bonded in a row then the cis position for the fourth carbon, C4 is as shown:

```
C1          C4
 \         /
  C2 —— C3
```

Another illustration of this can be seen in Figure 1.6 (p.4 4). Also see *trans*.

Cold Drawing The process of forming a polymer below its melting point whereby it is shaped in dies. This is best accomplished under conditions when the tensile test for the polymer displays necking. The result of drawing will be the alignment of the polymeric chains in the direction of draw. This will increase the elastic modulus and tensile strength in the drawing direction but will reduce these properties in the other directions.

Condensation A polymerizing reaction that results in a product which itself is further capable of reaction. Usually in condensation reactions a small molecule such as water is split off and the remainder of the molecule reacts with other similar molecules. This term is used to define the chemical reaction process by which a category of polymers is made. The final structure of the polymer determines its properties so the exact nature of the chemical reactions needed to produce it are not of the utmost importance for property prediction. However, by-products of intermediate reactions can affect the properties of the final product if these remain in the polymer. When this happens, the process of chemical reactions used to make a product is of importance in predicting its properties.

Conduction Band The next energy level beyond the normal or unexcited energy level occupied by an electron in an atom. When many atoms are combined this level broadens into a band of energy. The conduction band is so named because electrons elevated to it are free to conduct, or transmit energy.

Conjugated Double Bond The alternation of a single and a double bond along the backbone of a polymer. A ring of atoms, such as the carbon ring, may be considered as a unit along a backbone, which, then, also displays a conjugated double bond. The bond is free to resonate so the actual position of the double bond is always changing. This mobility of the bond allows for conduction of energy easily down the backbone.

Copolymer A polymeric material made up of two or more different types of monomers. They can react to form a polymer in a variety of different ways. The most usual is a random arrangement of monomers along the backbone. See *graft polymer.*

Corrosion The deterioration of a polymer by chemical or electrochemical reaction with its environment. Oxygen in the air is the single most corrosive element in a normal environment. The presence of ultraviolet light accelerates the reaction of oxygen with the polymer.

Coupling Agent A chemical used to improve the bond between a polymer and its reinforcing particles. Silane coupling agents are widely used between organic polymers and ceramic additives.

Covalent Bond A type of chemical bonding in which the participating atoms all contribute electrons in a shared manner.

Crazing Surface cracking occurring under applied stress or environmental attack. This is a time-dependent development that always worsens. If the cracking is extensive enough to be identified without instruments it is called crazing. When the cracks are so small that they blend together visually then it is called frosting. Crazes are partially filled with polymeric material that can still carry some load.

Creep The gradual deformation of a polymer under a load. The load may be an applied one or simply its own weight. Most polymers above the *glass transition temperature* are subject to extensive creep unless they are cross-linked, crystalline or otherwise reinforced.

Cross-Linking The chemical bonding, often with the addition of another atom, of one polymeric chain to another. These links are bonds between the chains that do not add to the direct extension of an existing polymer backbone. They do have the effect of changing the properties in a similar manner to increasing the molecular weight of each polymeric chain.

Crystalline A polymer will become crystalline when it is below its melting point, and it arranges itself into a regular three-dimensional array with a small repeating unit. The rate of crystallization will be dependent on the deviation of the melt from the melting point with the maximum rate usually occurring at a temperature equal to nine-tenths of the melting point on the absolute temperature scale.

Crystalline Melting Transition Temperature As a crystalline polymer approaches its melting point the crystalline structure becomes so mobile that it can be easily distorted by an external force. At this temperature, or temperature range, the mechanical strength of the material falls off precipitously. Usually this temperature is about 10°C below the melting point of the crystal.

Decompose The process by which a solid material breaks down directly into atoms or groups of atoms that are disassociated and energetic enough to be gaseous. Most polymers decompose rather than truly melt. The decomposition process begins long before the temperature usually assigned to it. It is a matter of rate or kinetics rather than absolutely a discontinuously abrupt change of state.

Defects, Crystalline A defect is any variation in the strict order of a crystalline lattice. Point defects in polymers are usually considered to be chain ends, short branches, and kinks. Chain ends and short branches are not usually included within the crystalline lamellae but left to dangle in the amorphous region. The kink, which is a unique crystalline defect of polymers, is brought about by a variation in bond orientation of the backbone. Since the kink allows for possible extra backbone atoms within a crystalline site, it can serve as a means of transport through the lamellae. Line dislocations of the edge and screw types are found in polymers as well as other types of crystals. Polymer crystals also display the usual planar defects of grain boundaries common to all crystalline types.

Density The mass per unit volume of a material usually expressed in grams per cubic centimeter or (kg/m^3).

Dielectric Strength The maximum voltage difference that can be sustained across a dielectric material before physical breakdown. Physical breakdown can be characterized both by deterioration of the material and by a catastrophic decrease in resistance. The average value for dielectric strength of a polymer is about 20 kV/mm. Thin sections, which have better values than thick sections, have been measured at 100 kV/mm.

Diffusion Rate The speed at which atoms, segments of backbone, or molecules move through one another. The diffusion rate is dependent on the vibrational energy of the type involved. It increases with increasing temperature and decreases with increasing pressure. Polymers are dependent on segment motion for mobility within the material, thus, the vibrational energy and the size of the mobile segment affect the diffusion rate.

Dipole An imbalance in the charge neutrality of a structure. A dipole can be an entire molecule, a cluster of atoms within the molecule, or a single atom in the molecule. Dipoles can be inherent in the structure of the molecule and thus a permanent part of the electrical character of that molecule. They can also be induced by an external force field and will return to neutrality when the force field is removed.

Ductility The characteristic of a material that absorbs extensive energy prior to fracture. Materials that exhibit this are able to be plastically deformed and are difficult to break.

Dye Particles of small size that are added to a polymer to give it a translucent coloration. Dyes are also used in conjunction with *pigments* to produce a wider variety of hues. They are also useful in aiding flow properties for extrusion processing, but they do limit the practical processing temperature.

Elastic Region That part of the tensile flow behavior in which the strain is linearly related to the stress, i.e., the material returns to its original dimensions upon the removal of the stress. In polymers, only cross-linked materials have an appreciable elastic region over a wide temperature range.

Elastomer Polymers that are able to withstand large deformation and return to their original shape when the deforming force is removed. This property appears when the backbone of the polymer can readily undergo torsional motions. The most effective elastomers are cross-linked.

Electret An electret is produced by heating an electrically insulating polymer to 50°C above its glass transition temperature and then subjecting it to an electric field of about 2.5 kV/mm for an hour. The resulting electret is then cooled below its glass transition temperature while the electric field is still in effect. These conditions produce ions in the additives and impurities of the polymer, which will take months or years to recombine, unless the material is heated above the glass transition temperature again.

Electroluminescence The emission of light by a material as a result of the application of an electric field.

Electrolytic Corrosion Fine, current-carrying wires of 0.02 cm or less are susceptible to corrosion at the positive junction when in direct contact with a polymer. This is particularly prevalent in direct current curcuits but may also occur in alternating current curcuits. Polymers that are polar enough to attract water are most likely to cause electrolytic corrosion.

Electrophotography Exposure of light-sensitive polymers to a light source to produce a temporary rearrangement of structure. Electrophotography has found use in imaging.

Elongation The extent to which a material can extend during tension testing. It is generally expressed as the ratio of final length to initial length. Polymers display the widest range of elongation of any material. Usually cross-linked polymers have limited elongation, whereas symmetric non-polar polymers generally elongate extensively.

ESD—Equivalent Spherical Diameter The diameter of a sphere having the same volume as that of a particle. This is used extensively with respect to additives. It gives a basis of comparison for irregularly shaped particles.

Expanded Plastic See *foam.*

Extender See *filler.*

Extraction (re: plasticizers) Plasticizers and host polymers are usually soluble in the same solvent, since an effective plasticizer must be soluble in its host. However, since the plasticizer is by its nature smaller and thus has a better penetrating ability than its host, it will also tend to be more soluble in all mutually soluble materials. Thus, when a plasticized polymer is exposed to a mutual solvent, the plasticizer is preferentially extracted by the solvent over the host polymer.

Exudation (re: plasticizers) The gradual escape of plasticizers through the surface of a polymer. The plasticizer continues to diffuse to the surface so the process is continuous and temperature dependent. The presence of the plasticizer on the surface often imparts a greasy feeling to the polymer and encourages the clinging of dirt to it. Exudation increases with decreasing *solubility* of the plasticizer.

Fiber An aligned polymer which has its principal strength along the aligned direction. Fibers can be on the microscopic level as part of the structure of a polymer or they can be so large as to be used as filaments in yarn production.

Fibril Very long connections between aligned lamellae usually radiating out from a center nucleus in a crystalline structure. A fibril may be up to 15,000 Å in length and 300 Å across. Also see *tie molecule* and Figure 1.21 (p. 17).

Filler An additive used to increase the volume of the polymer. These are almost always very cheap materials that provide no other positive benefit to the product.

Flexibility The ability of a material to be deformed without fracture. Polymers may have very high values of flexibility and this fact is often used to advantage in design.

Flexural Strength The maximum extreme stress that a material can withstand in bending prior to failure. Flexural strength, like flexibility, is often very high in polymers and allows designs otherwise impossible.

Flow Temperature The transition temperature from rubber to *viscofluid* behavior. Most high-molecular-weight polymers do not display this because they go directly from the rubber stage into degradation. If a polymer does display a viscofluid region, it can often be used to advantage in processing.

Fluorescence The emission of visible light by a polymer upon exposure to a source of energy. Electrons of a polymer are elevated into the conduction band if there is an energy source of the proper wavelength. If on removal of that energy source the electrons quickly return to their previous energy level, then they do so by giving off a photon of energy. If the wavelength of the emitted photon of energy is in the visible range, the polymer is said to exhibit fluorescence.

Foam A two-phase mixture of polymer and gas in which the gas is usually the foaming agent. The amount of polymer present can vary from as low as 1% up to 100%. Usually the normal polymer properties will be the properties of the foam after correction has been made for the reduced volume for extensive properties.

Foaming Agent The gaseous material which causes a polymer to become a foam. See *foam.*

Free Volume The difference in volume between that of the material at a particular temperature and that which it would occupy at absolute zero. This parameter is used to characterize different states for polymers. It is also used in many mobility considerations.

Fresnell Effect As stress is applied to a polymer with an additive, separation often occurs causing a void. The void index of refraction is different from the surrounding material and a whiter appearance results. Because the size of the particle determines the size of the void, whitening will appear more readily in a polymer filled with large particles. Thus the larger the particle, the smaller the stress required for the whitening to appear when the volume fractions are the same.

Frosting See *crazing.*

Gauche The most common form for the backbone chain to exhibit is the trans but some bonds will be in the gauche position at normal temperatures. The gauche allows the backbone to twist out of its normal planar position so that a chain can take on a more three dimensional aspect. This usually occurs around large pendants like carbon rings. The gauche position forms a 120° angle with the chain.

See Figure 1.7 (p. 5) for an example of the gauche position. Also see *trans*.

Glass Transition Temperature The temperature at which a glass transforms into a rubber for the amorphous part of the polymer. For a polymer which displays some crystallinity and is *isotactic* the ratio of glass transition temperature to melting temperature on an absolute scale is about 0.5. When a polymer displays crystallinity and is *atactic* this ratio is about 0.7. The value of the glass transition temperature for a particular polymer is sensitive to molecular weight, branching, crystallinity, and traces of solvents from processing.

Glassy State A thermodynamically metastable phase within the liquid region of a polymer. It is characterized by short-range order and short time stiffness. Long-term stiffness decreases with increasing temperature until the glass transition temperature is reached.

Graft Polymer A type of copolymerization where polymeric chains made up of monomers differing from all or most of the monomers comprising the backbone of the main polymer are attached as side chains. The backbone may itself be a copolymer.

Heat Capacity A loosely defined term often used interchangeably with specific heat. See *specific heat*.

Hybridization The blending of a covalent bond to accommodate overall bonding strength of a molecule. The best example of this is the equivalent bonds formed within the carbon regular tetrahedron of a diamond.

Hydrogen Bond Stronger than the average van der Waals bond, although often classified with it, it is still a secondary bond. When hydrogen bonds to a larger atom, such as carbon, the lone electron it contributes spends most of its time orbiting near the larger atom, although it may still be classified as a covalent primary bond. As such, it leaves the hydrogen nucleus, a single proton, partially exposed most of the time. The hydrogen nucleus has a unit positive charge and is capable of attracting negatively charged dipoles or radicals. When this happens a "hydrogen bridge" is formed. Such bonding may occur between segments within a single polymeric chain or between polymeric chains. The hydrogen bond is dependent on spatial separation, as are all bonds. When these are favorable it can be about 100 times stronger than just the van der Waals bond. Thus, the hydrogen bond can serve as a very important interchain bond. Since most polymers contain hydrogen, its possibility of occurring can be quite high. Structural modifications are often done in processing to further enhance this type of bonding.

Index of Refraction The ratio of the velocity of light in a vacuum to its velocity in a polymer. In the case of polymers, almost all of the measured values are taken in air rather than in vacuum. As light travels from the air into a polymer its velocity is reduced and its angular direction is shifted toward the vertical in accordance with Snell's law.

Ionic Bonding The chemical bonding of two or more atoms such that the valence electrons are predominately associated with one type of elemental atom. This is not a very common form of primary bonding in polymers.

Isomers Materials which are made up of the same elements but of different arrangement. The difference in arrangement causes the materials to display different measurable property values. Polymers are often very capable of isomerism because the angular relationship of the backbone can vary. See *cis* and *trans*.

Isotactic The stereoregularity of a polymer can affect its properties, notably its ability to crystallize. If the side atoms of pendant groups of a polymer are only located on one side relative to the backbone then it is said to be isotactic. Since this is a regular structure, it is often favorable for crystallization. Schematically it appears as:

Jump Frequency The rate at which a polymeric chain segment will move into an available space. The jump frequency is usually tabulated in terms of jumps per second.

Lamellar Structure The single crystal structure for a polymer under most circumstances. See Figure 1.19 (p. 16).

Linear Polymer A polymer without branching or cross-linking.

Liquid State The phase in which short-range order is displayed but there is no stiffness. A polymer which flows freely under its own weight is considered to be liquid. Strictly speaking, everything which is not crystalline and not obviously a gas falls into the liquid category. However, since most polymers do have significant mechanical strength even when non-crystalline, the practical definition of a liquid is better based on stiffness than structure. Most high-molecular-weight polymers go directly from the "solid" phase to the gaseous without displaying a true liquid phase.

Loss Factor The product of the *dielectric constant* and the *power factor*. The loss factor reflects the amount of signal absorption which has taken place and is usually a low value for polymers.

Luminescence The general term for visible light emission after energy input, including both fluorescence and phosphorescence.

Macromolecule An alternate name for a polymer, usually of the structural type; giant molecule.

Microballoons Hollow spheres which are used to produce what are effectively reinforced foams. The microballoons are usually made of glass, which serves as a strengthening and stiffening additive to the polymer. In addition to improved mechanical properties, the resulting polymer has reduced density and is useful for floatation.

Migration (re: plasticizer) If a plasticized polymer is put into contact with a non-plasticized polymer, the plasticizer has a tendency to diffuse or migrate into the unplasticized polymer until an equilibrium is reached. The relative solubilities of the plasticizer in the two polymers will limit the amount of migration. See *inter-diffusion*.

Modulus of Elasticity The ratio of *stress* to *strain* in tension at strains that are sufficiently small that the material will return to its original length when the stress is removed. For cross-linked and rigid polymers this value has useful meaning but for a large number of polymers it is mainly a traditional measurement. See *tensile modulus*.

Molecular Weight The total weight of a polymeric chain. Usually it is given for individual molecules, which have weights that are quite small.

Molecule A group of atoms that are bound together by either ionic or covalent bonding. It is the basic unit of the polymer since each chain is a molecule.

Monomer The small molecule that is reacted to produce a polymer. Only the portion which is actually used in the chemical reaction is considered to be the monomer. For example, polyethylene is made up of repeating monomers of ethylene but only the:

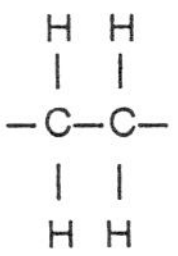

is considered as the monomer

rather than the starting molecule:

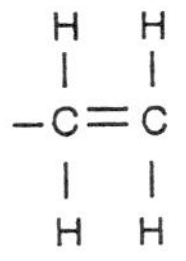

Mullens Effect Filled elastomers display a characteristic softening when stressed cyclically. This effect can be reduced by allowing the elastomer a long relaxation time between cycles but does not seem to be completely recoverable.

Necking The local reduction in cross-section of a material while it is undergoing unconstrained deformation. In amorphous polymers, this usually occurs just below the glass transition temperature. For polymers with appreciable crystallinity, the temperature of necking is usually just above the glass transition temperature. The appearance of necking is also sensitive to the rate of loading and to prior orientation of the polymer. See Figures 4.2 and 4.3 (p. 44).

Open Cell When the gaseous phase or "cell" of a foam is not completely surrounded by the polymeric phase. Closed-cell means that the polymeric phase is continuous. Most commercial foams are at least partially of the open cell type. This has disadvantages if the foaming agent is to be retained in the final product or if the foam is intended for floatation.

Orbital The path that an electron takes around the nucleus of the atom to which it is bound. In the case of most polymers, only the *s* and *p* orbitals are of interest. See *s orbital* and *p orbital*.

Organic Material which was originally formed by a living organism. Most polymers have an organic origin and are thus susceptible to attack by organisms that feed on organic matter.

***p* Orbital** A dumbell shaped region around the nucleus of an atom in which electrons of distinct energy values can be found. Up to two electrons can occupy a *p* orbital at any one time. In an energy level there are one to three *p* orbitals at an energy level, which are usually at right angles to each other. See Figure 1.1 (p. 2).

Particle Radiation Bombardment by sub-atomic particles which may include X- and gamma-radiation. Thin sections of polymers allow full passage of alpha-, beta-, gamma-, and X-radiation. Thicker sections of polymers will still allow full passage of gamma- and X-radiation, provided the polymer does not contain additives that reflect or absorb the radiation. Neutrons are generally absorbed by all thicknesses of polymers. Polymers have been used as screening agent for radiation. Radiation is also a means of producing structural modification in a polymer.

Pendant Group A cluster of atoms chemically bonded to the backbone of a polymer. Usually this cluster is large and may contain unoccupied space, such as a carbon ring. See Figure 1.9 (p. 7).

Permeability The amount of a second material, usually a fluid, that can be moved through a unit area of a host material in a given time. The value of this property is very dependent on the pressure of the fluid moving through the polymeric host. The greater the pressure of the fluid, the greater the driving force through the polymer will be. Thus, flow through the polymer will proceed more quickly. Also, increasing the temperature of the polymer will increase the permeability of the fluid since the polymeric free space increases with increasing temperature.

Phosphorescence The emission of light that occurs under the same conditions as fluorescence, but the recombination process happens over a period of time which results in a decaying phenomenon.

Photocrosslinking Light-sensitive polymers can be selectively exposed to ultraviolet light causing the exposed segments to cross-link and become hard and relatively insoluble. The soluble unexposed remaining polymer can then be completely dissolved, leaving an image. See *photopolymerization*.

Photon A quantum of electromagnetic energy. The amount of energy required to move an electron from one energy level to the next within an atom will require an integral number of photons. The quantity of this energy or its wavelength helps define characteristics of a particular element. Since polymers are mainly made up of a few elements, the photons associated with polymers fall into a very narrow band.

Photopolymerization Light-sensitive monomers can undergo selective polymerization upon exposure to a light source. The extreme accuracy of the process allows for very intricate miniature patterns to be produced in a subsequently structurally strong polymer. This has found extensive application in the electronics industry. See *photocrosslinking*.

Piezoelectric Polymers which generate a current during or after deformation. This is not the conventional concept of a piezoelectric material, since the concept was originally considered to be possible only in crystals. Many polymers, however, have dipoles induced when their bonds are deformed by mechanical force. The recombination of these dipoles produces a small de-polarization current. The recombination may take place immediately or over a long period of time depending on the mobility of the dipoles. Once the dipoles have recombined, the polymer will have returned to electrical neutrality but will not necessarily remain undeformed.

Pigment A large-sized particle added to a polymer to give it opaque color. Both organic and inorganic pigments are used. The most common pigments are titanium salts and carbon-black.

Plastic The word is derived from the Greek word, *plastikos,* meaning easily shaped. It is now used as a general term for all macromolecular materials.

Plasticizer A liquid or solid additive used to make a polymer more fluid. It must have at least limited solubility in a polymer to be effective. This usually means that it must be chemically similar to its host polymer.

Plasticizer Efficiency The amount of fluidity produced in a host polymer versus the weight percentage of plasticizer needed to produce that fluidity. Since it is desirable to limit the volume of plasticizer used, this is a helpful means of evaluation.

Polymer The word is derived from the Greek words for many, *poly,* and small part, *mer.* Polymers are indeed made up of many small parts or repeating units of atoms chemically bonded to form a molecule.

Polymer Blend or Polyblend A mixture of two or more polymers which are combined by mechanical means only. Such polymers are not linked together by chemical bonds.

Power-Factor The ratio of measurable power to the voltage-current product. This is usually a high value for polymers, since they are non-conductive.

Quench The process of quickly cooling a material to preserve as much as possible of its high temperature structure. In a polymer, it may be appropriate to quench to a lower temperature within the rubbery region. All quenching to the glassy region will be sufficient to maintain the high temperature structure.

Radical A chemically bonded group of atoms which has a net electrical charge associated with it. For example, the oxygen-hydrogen combination has a negative charge of one. If the polymer containing this radical is dissolved or heated to sublimation, then the oxygen-hydrogen combination would likely remain bound as a unit, although no longer bound to the rest of the polymer.

Reaction Site The place on the atom or group of atoms where a chemical reaction will take place.

Reflected Light Light that strikes a surface and is redirected away from the surface, obeying an angular relationship with the incident beam. The angular relationship is that the angle of incidence of the incoming light beam be equal to the angle of the outgoing reflected beam. See Figure 3.8. Since most polymers have a smooth surface the amount of reflected light is quite high resulting in a shiny appearance.

Reinforcement Additive A particle added to a polymer for the purpose of increasing the tensile strength of the material. The optimum diameter of a ceramic reinforcement additive is about 10 nm.

Resonating Multiple Bonds When an atom is bonded with a multiple bond on one side and a lesser number of bonds on the other side one or more of the "extra" bonds may be able to switch positions. The overall effect is sharing of the extra bonds by the molecule. Along with this sharing comes a reduction in overall bond length due to a reduction in repulsive forces.

Reverberation The multiple reflections of sound waves. Polymers are often used in designs in which flat metal surfaces would cause a great deal of reverberation, because polymers are very low in sound reflectivity.

Rubber See *elastomer.*

***s* Orbital** A spherically shaped region around the nucleus of an atom in which electrons of distinct energy values can be found. See Figure 1.1 (p. 2).

Saturated Bond A polymer is said to have saturated bonding when the backbone only contains single bonds. Any addition of double or triple bonding to the backbone causes it to be termed unsaturated.

Scission The breaking of chemical bonds irreversibly. Usually this term is used in the breaking up of the backbone of a polymeric chain.

Self-Diffusion The random diffusion of segments of a polymeric chain within a polymer matrix. If the chains were of a different type, then the mixing of these types would be diffusion and proceed in a specified direction rather than randomly.

Shell Electrons are arranged within an atom by orbitals, and orbitals are located within shells. The outer limit of an orbital will fall within its shell. The shell is a spatial designation and does not have a specific energy level associated with it. The primary reason to be concerned with shells in polymers is that the shape of the outer shell influences the shape and orientation of the bonds that form.

Side Chains See *branched molecules.*

Soluble When two materials are sufficiently compatible that they can intermix without chemical reaction. Solubility has a range of values from completely insoluble to mixable in any proportion. Most polymers of similar atomistic composition are at least slightly soluble in one another. This term is not usually used in conjunction with crystalline materials.

Specific Heat The amount of energy in calories needed to raise one gram of material 1 °C. If this value is multiplied by the density of the material, the result will be the *heat capacity* on the basis of volume. Volumetric heat capacity is often a more useful value for polymers than that based on weight.

Spherulites A crystalline, super-molecular structure which can range in size from a fraction of a micron to over a centimeter in diameter. As the name suggests, it is spherical in shape and usually results when a polymer that is not in solution crystallizes. See Figure 1.21 (p. 17).

Static Charge The build-up of an electrical charge on a polymer, usually at the surface. This is a common problem with polymers since they are unable to dissipate electrical charges effectively through conduction. The smoothness, temperature, and additives all affect static charge build-up.

Sponge Plastic See *foam.*

Strain Generally considered to be the linear strain or the change in length in the direction of an applied stress divided by the original unstressed length. Measuring this isothermally is often very difficult for a polymer since heat build-up often occurs during stressing.

Strain Hardening An increase in strength accompanying strain. This term may be said to apply to the last segment of a tensile test graph when a polymer shows an increase in sustained stress before failure. This is usually a result of a hardening mechanism, such as chain alignment, in the direction of the applied force.

Strength The maximum load per unit area that a material may be considered to be capable of supporting. This property is very sensitive to the testing procedure used and consequently difficult to compare.

Stress The applied force or load divided by the original cross-sectional area carrying the force. Since *necking* can be extensive in a polymer, this value is often deceptively low.

Stress Whitening See *Fresnell effect.*

Structural Polymer One which is able to sustain a load in at least two directions. Usually fibers are not included in this category, although they have been applied in functions for which structural polymers were previously used.

Sublime The process by which a solid material breaks down directly into atoms or groups of atoms that are disassociated and energetic enough to be gaseous. Most polymers sublime rather than truly melt. The sublimation process begins long before the temperature usually assigned to it. It is a matter of rate or kinetics rather than absolutely a discontinuously abrupt change of state.

Swelling The dissolving of a solvent into a cross-linked polymer. The cross-linked polymer will elastically stretch to accommodate the solvent, thereby producing a gel in the case of extreme solubility. Swelling is a common problem when rubbers are used in conjunction with greases.

Syntactic An arrangement in which the side atoms or groups alternate. Schematically it would look like:

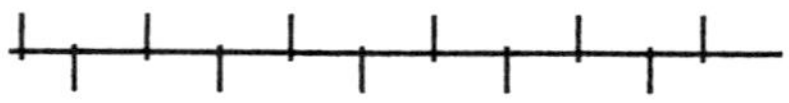

Tensile Modulus In thermosetting polymers this term is interchangeable with *modulus of elasticity* because the cross-linking produces a material with an elastic region which corresponds to its entire initial linear region. However, for a thermoplastic, the elastic region extends to about 2% strain with a reasonably linear region of the tensile curve continuing well beyond the actual elastic limit; the entire linear region is used for the determination of the tensile modulus. Thus, for thermoplastics the more usually reported value is tensile modulus and it is becoming the term of choice for most polymers.

Tensile Strength The maximum load that a polymer can withstand prior to failure divided by the original cross-sectional area supporting the load. Since many polymers *neck* to a large extent, this value may be less than would be expected.

Tensile Test A procedure in which a polymeric sample is loaded in tension for the determination of its modulus, strength, elongation and toughness. Although this is a standard testing procedure, not all tests on polymers are carried out under identical conditions, thereby creating confusion in data interpretation.

Thermal Conductivity The amount of heat energy a polymer can transmit through its bulk as a function of time. The average amorphous polymer as a thermal conductivity of about 0.0004 cal/cm-sec-deg (Celsius), whereas a crystalline polymer would have about twice that value.

Thermal Degradation The process through which a polymer is decomposed by energy supplied in the form of heat, breaking its chemical bonds. Minor bond breakage can occur at any temperature, so a characteristic temperature for thermal stability is defined. This is the temperature at which one half of the weight of the polymer is lost in one-half hour and is designated by T 1/2. The test is conducted in a vacuum to prevent the recombination of other elements, such as oxygen, to distort the weighing.

Thermal Expansion The ratio of the change in volume per degree Celsius to the volume at zero Kelvin. Often the thermal expansion in one direction is the only design concern. In that case, the volumetric terms can be replaced by linear terms. The coefficient of linear expansion is approximately equal to one third the coefficient of volume expansion. For most polymers, the expansion due to increasing temperature is much less than the change that is produced by the glass transition. Thus, thermal expansion data are often listed only as values just below and just above the glass transition temperature rather than as a function of temperature.

Thermal Expansion Empirical Relationships The change in thermal expansion due to the glass transition temperature is about 0.0005°C. When this value is multiplied by the glass transition temperature the result is about 0.115. Thus, the expected glass transition temperature, if both of these empirical relationsahips are obeyed, is 230 Kelvin. If the thermal expansion above the glass transition temperature is multiplied by the glass transition temperature, the result will be approximately 0.16. For crystalline polymers, the product of the thermal expansion and the melting temperature is about 0.11. The thermal expansion of a crystal is approximately that of the polymer below the glass transition temperature.

Thermoluminescence The emission of light as a result of heat input.

Thermoplastic A polymer that softens as it is heated without a specific softening point. Amorphous thermoplastics go from the glass transition temperature through to a liquid, or gas, with ever increasing softness. Crystalline thermoplastics have a better defined melting point but still exhibit increasing softening from the glass transition temperature to the melting point.

Thermosetting A polymer that does not display a softening point. This type of polymer is produced by starting with a monomer that will form a highly cross-linked matrix unable to respond to elevated temperatures in any way except sublimation or char.

Tie Molecules When the same polymeric chain grows into two different lamellae, the short section of chain between the lamellae serves as a tie between them. Tie molecules unify and reinforce the structure of the crystal. See Figure 1.18 (p. 15).

Tough Capable of absorbing substantial energy at high rates prior to failure. Most polymers are tough at ambient temperature.

Toughness The quality of being *tough*.

Tracking A combination of dust and moisture can produce an electrical path on the surface of a polymer being used as a dielectric. If the path deteriorates physically, then the material is said to exhibit tracking. Once tracking has occurred, it will be in this area that the dielectric breakdown will occur. With concentration of power along the narrow tracking path, the polymer may become so overheated locally as to ignite.

Trans The backbone of a carbon or silicone based polymer is not straight but zig zags along at about a 110° angle. Since the bond is free to rotate about the angle in space, all positions are theoretically possible but only a few are energetically favorable. The most favorable positions are the trans followed by the gauche. The cis is the least favorable position. All other positions are considered energetically unfavorable. Whereas the cis represents an angular relationship of 180° to the backbone, the trans is that of 0°. This is shown in Figure 1.5 (p. 4) and in the following schematic where atom C4 is in the trans position:

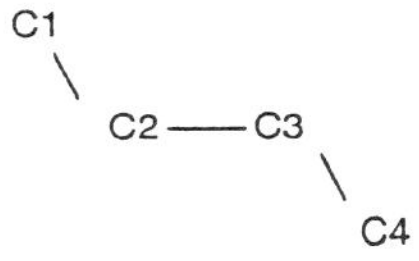

Triboluminescence Luminescence resulting from mechanical deformation.

Unit Cell The smallest arrangement of atoms in three-dimensional space that can be repeated in all directions to form a crystalline solid. In the case of polymers, the bonding along the backbone limits the arrangements possible but does not interfere with the normal definition of this term.

van der Waals Bond A form of secondary bonding that may be considered to include hydrogen bonding as well. There is no covalent or ionic bonding involved, so the van der Waals bond acts between separate polymeric chains that are in close proximity to each other. It is really a combination of all minor attraction forces, so it includes mass attraction and dipole-dipole attraction.

Visco-Fluid State The state that exists above the rubber state in polymers that do not sublime prior to reaching it. It is exhibited by amorphous thermoplastics. This state has more fluidity than the rubber state but is not sufficiently fluid to be considered a true liquid.

Volatility (re: plasticizers) Plasticizers are usually low molecular weight organic polymers that are subject to evaporation. The addition of thermal aging to polymer processing makes their evaporation that much more favorable and removes the plasticizer from the polymer much more quickly.

Vulcanization The process by which sulfur is added to natural rubber to produce a cross-linked polymer.

Yield Strength The stress at which a material exhibits a specified deviation from the proportionality between approximately linear stress and strain. Often the 0.2% offset yield strength is tabulated, such that the yield strength represents a plastic strain of 0.002 in uniaxial tension. For many polymers, there is no real yield strength, as plastic deformation commences immediately.

Young's Modulus See *modulus of elasticity.*

BIBLIOGRAPHY

Agranoff, J., ed. *Modern Plastics Encyclopedia,* Vol. 61, No. 10A, NY:McGraw-Hill (1984).

Aisenberg and King. *Ion-Containing Polymers,* NY:Academic Press (1977).

Allcock and Lampe. *Contemporary Polymer Chemistry,* Englewood Cliffs, NJ:Prentice-Hall (1981).

Annual Book of Standards, 1984, American Society for Testing Materials, Phila., Pa. (1984)

Baer, Eric. *Engineering Design for Plastics,* NY:Reinhold (1964).

Baijal, Mahendra. *Plastics Polymer Science and Technology,* NY:John Wiley and Sons (1982).

Baird, R. J. and D. T. Baird. *Industrial Plastics,* South Holland, IL:The Goodheart-Willcox Co., Inc. (1986).

Bely, Sviridenok, Petrokovets and Savkin. *Friction and Wear in Polymer-Based Materials,* NY:Pergamon (1982).

Benjamin, B. *Structural Design with Plastics,* NY:Van Nostrand Reinhold (1982).

Bikales, Norbert, ed. *Mechanical Properties of Polymers,* NY:Wiley-Interscience (1971).

Billmeyer, Fred, Jr. *Textbook of Polymer Chemistry,* NY: Interscience (1957).

Birley, A. W. and M. J. Scott. *Plastics Materials Properties and Applications,* London, England:Leonard Hill (1982).

Blythe, A. *Electrical Properties of Polymers,* London: Cambridge University Press (1979).

Bolker, Henry. *Natural and Synthetic Polymers,* NY:Marcel Dekker (1974).

Boretos, John. *Concise Guide to Biomedical Polymers,* Springfield, IL:Charles C. Thomas (1973).

Brown, R. *Design and Manufacture of Plastic Parts,* NY:John Wiley and Sons (1980).

Bruins, Paul, ed. *Plasticizer Technology,* NY:Reinhold (1965).

Bruins, Paul, ed. *New Polymeric Materials,* NY:Interscience (1969).

Brydson, J. A. *Plastics Materials,* Princeton, NJ:D. Van Nostrand (1966).

Bueche, F. *Physical Properties of Polymers,* NY:Interscience (1962).

Cantow et al., ed. *Advances in Polymer Science,* NY: Springer-Verlag (1980).

Carraher, Sheats and Pittman, ed. *Organometallic Polymers,* NY:Academic Press (1978).

Cassidy, Patrick. *Thermally Stable Polymers,* NY:Marcel Dekker (1980).

Chapiro, Adolphe. *Radiation Chemistry of Polymeric Systems,* NY:Interscience (1962).

Clark, Reed and Hartwig, ed. *Nonmetallic Materials and Composites at Low Temperatures,* NY:Plenum (1979).

Conley, R. T., ed. *Thermal Stability of Polymers,* NY: Marcel Dekker, Inc. (1970).

Crawford, R. J. *Plastics and Rubbers Engineering Design and Applications,* London:Mechanical Engineering Pub. Ltd. (1985).

DiBenedetto. *The Structure and Properties of Materials,* NY:McGraw-Hill (1967).

Elias, Hans-Georg. *New Commercial Polymers 1969–1975,* NY:Gordon and Breach (1977).

Engel, Klingele, Ehrenstein and Schaper. *An Atlas of Polymer Damage,* Englewood Cliffs, NJ:Prentice-Hall (1981).

Ferry, J. D. *Viscoelastic Properties of Polymers,* NY:John Wiley & Sons, Inc. (1970).

Flory, Paul. *Principles of Polymer Chemistry,* Ithaca, NY: Cornell University Press (1953).

Frisch and Sanders, ed. *Plastics Foams,* NY:Marcel Dekker (1972).

Golud, Parker, ed. *Polymeric Materials for Unusual Service Conditions,* NY:John Wiley and Sons (1973).

Hall, Christopher. *Polymer Materials,* NY:John Wiley and Sons (1981).

Hanks, Richard. *Materials Engineering Science,* NY: Harcourt, Brace and World (1970).

Hertzberg and Manson. *Fatigue of Engineering Plastics,* NY:Academic Press (1980).

Kamal, Musa, ed. *Weatherability of Plastic Materials;* NY:Interscience (1967).

Katz and Milewski, ed. *Handbook of Fillers and Reinforcements for Plastics,* NY:Van Nostrand (1978).

Kaufman and Falcetta, ed. *Introduction to Polymer Science and Technology,* NY:John Wiley and Sons (1977).

Kieffer, William. *Chemistry: A Cultural Approach,* NY: Harper and Row (1971).

Landrock, A. H. *Handbook of Plastics Flammability and Combustion Toxicology,* Park Ridge, NJ:Noyes Pub. (1983).

Lazan, B. J. *Damping of Materials and Members in Structural Mechanics,* NY:Pergamon Press (1968).

Makhlis, F. *Radiation Physics and Chemistry of Polymers,* NY:John Wiley and Sons (1975).

Manson and Sperling. *Polymer Blends and Composites,* NY:Plenum (1976).

Mascia, L. *The Role of Additives in Plastics, NY:John Wiley and Sons (1974).*

Mascia, L. Thermoplastics: Materials Engineering, NY: Applied Science (1982).

Miller, M. *The Structure of Polymers,* NY:Reinhold (1966).

Moore, G. R. and D. E. Kline. *Properties and Processing of Polymers for Engineers,* Englewood Cliffs, NJ:Prentice-Hall (1984).

Nielsen, Lawrence. *Mechanical Properties of Polymers,* NY:Reinhold (1962).

Palin, G. *Plastics for Engineers,* London:Pergamon (1967).

Pauling, Linus. *The Nature of the Chemical Bond,* Ithaca, NY:Cornell Univ. Press (1960).

Perepechko, I. *Low-Temperature Properties of Polymers,* Elmsford, NY:Pergamon Press (1980).

Powell, Peter. *Engineering with Polymers,* NY:Chapman and Hall (1983).

Ranby, B., ed. *Long-Term Properties of Polymers and Polymeric Materials,* NY:John Wiley and Sons Inc. (1979).

Rigney, David, ed. *Fundamentals of Friction and Wear of Materials,* Metals Park, Ohio:American Society for Metals (1981).

Ritchie, P., ed. *Plasticisers, Stabilisers and Fillers,* London:Iliffe Books (1972).

Rodriguez, Ferdinand. *Principles of Polymer Systems,* NY:McGraw-Hill (1970).

Samuels, Robert. *Structured Polymer Properties,* NY:John Wiley and Sons (1974).

Schnabel, W. *Polymer Degradation,* NY:Macmillian (1981).

Schultz, Jerold. *Polymer Materials Science,* Englewood Cliffs, NJ:Prentice-Hall (1974).

Seminar of ASM, *Polymeric Materials,* Metals Park, Ohio:American Society for Metals (1975).

Seymour, Raymond, ed. *Additives for Plastics,* NY:Academic Press (1978).

Seymour, Raymond. *Plastics versus Corrosives,* NY:John Wiley and Sons (1982).

Smith, Paul. *Plastics as Metal Replacements,* Broseley, England:Scientific Publishers (1968).

Tobolsky, Arthur. *Properties and Structure of Polymers,* NY:John Wiley and Sons (1960).

Ulrich, Henri. *Introduction of Industrial Polymers,* NY: Macmillian (1982).

Van Krevelen and Hoftyzer. *Properties of Polymers,* NY: Elsevier (1976).

Ward, I., ed. *Structure and Properties of Orientated Polymers,* NY:John Wiley and Sons (1975).

Whittington, L. *Whittington's Dictionary of Plastics,* Lancaster, PA:Technomic (1978).

Williams, David. *Polymer Science and Engineering,* Englewood Cliffs, NJ:Prentice-Hall (1971).

Williams, J. G. *Fracture Mechanics of Polymers,* NY:John Wiley and Sons (1984).

Winding and Hasche. *Plastics: Theory and Practice,* NY: McGraw-Hill (1947).

Winding and Hiatt. *Polymeric Materials,* NY:McGraw-Hill (1961).

Wyatt and Dew-Hughes. *Metals Ceramics and Polymers,* London:Cambridge University Press (1974).

Young, Robert. *Introduction to Polymers,* NY:Chapman and Hall (1981).

INDEX